Philip H. Gosse

The Wonders of the Great Deep

The Physical, Animal, Geological and Vegetable Curiosities of the Ocean

Philip H. Gosse

The Wonders of the Great Deep
The Physical, Animal, Geological and Vegetable Curiosities of the Ocean

ISBN/EAN: 9783337236427

Printed in Europe, USA, Canada, Australia, Japan

Cover: Foto ©berggeist007 / pixelio.de

More available books at **www.hansebooks.com**

THE WONDERS

OF

THE GREAT DEEP;

OR THE

PHYSICAL, ANIMAL, GEOLOGICAL, AND VEGETABLE

CURIOSITIES OF THE OCEAN.

BY

P. H. GOSSE,

AUTHOR OF "ROMANCE OF NATURAL HISTORY."

———

NEW YORK:
JOHN W. LOVELL COMPANY,
150 WORTH STREET.

THE WHALE FISHERY.

PREFACE.

In the following pages, the Author has endeavoured to describe, with some minuteness of detail, a few of the many objects of interest more or less directly connected with the Sea, and especially to lead youthful readers to associate with the phenomena of Nature, habitual thoughts of God. A subject so vast as the Ocean might be viewed in a variety of aspects, all of them more or less instructive: the one which has been chosen is that in which it presents itself to the mind of a naturalist, desirous of viewing the Almighty Creator in His works. The selections are made chiefly from marine botany, zoology, meteorology, the fisheries, the varying aspects of island and coast scenery, incidents of navigation, &c., arranged (if such a word be not inapplicable) in the order of geographical distribution; as they might be supposed to present themselves to the notice of an observant voyager.

It may be thought that the Author has touched too frequently, or dwelt with too great prolixity, on objects minute in themselves, and

by the generality of persons considered insignificant and unworthy
of regard.   If apology for this be necessary, he presents it in the
words of Samuel Purchas:—"Nicostratus in Ælian, finding a
curious piece of wood, and being wondered at by one, and asked
what pleasure he could take to stand, as he did, still gazing on the
picture, answered, 'Hadst thou mine eyes, my friend, thou wouldst
not wonder, but rather be ravished, as I am, at the inimitable art
of this rare and admirable piece.'   I am sure no picture can express so much wonder and excellency as the smallest insect, but we
want Nicostratus his eyes to behold them.

"And the praise of God's wisdom and power lies asleep and dead
in every creature, until man actuate and enliven it.   I cannot,
therefore, altogether conceive it unworthy of the greatest mortals
to contemplate the miracles of Nature; and that as they are more
visible in the smallest and most contemptible creatures (for there
most lively do they express the infinite power and wisdom of the
great Creator), and erect and draw the minds of the most intelligent
to the first and prime Cause of all things; teaching them as the
power, so the presence, of the Deity in the smallest insects."

CONTENTS.

(7)

III. THE ARCTIC SEAS.

IV. THE ATLANTIC OCEAN.

ILLUSTRATIONS.

THE OCEAN.

WHO ever gazed upon the broad sea without emotion? Whether seen in stern majesty, hoary with the tempest, rolling its giant waves upon the rocks, and dashing with resistless fury some gallant bark on an iron-bound coast; or sleeping beneath the silver moon, its broad bosom broken but by a gentle ripple, just enough to reflect a long line of light, a path of gold upon a pavement of sapphire; who has looked upon the sea without feeling that it has power

"To stir the soul with thoughts profound?"

Perhaps there is no earthly object, not even the cloud-cleaving mountains of an alpine country, so sublime as the sea in its severe and naked simplicity. Standing on some promontory whence the eye roams far out upon the unbounded ocean, the soul expands, and we conceive a nobler idea of the majesty of that God, who holdeth "the waters in the hollow of His hand." But it is only when on a long voyage, climbing day after day to the giddy elevation of the

B   (13)

mast-head, one still discerns nothing in the wide circumference but the same boundless waste of waters,
that the mind grasps anything approaching an adequate idea of the grandeur of the Ocean. There is
a certain indefiniteness and mystery connected with
it in various aspects that gives it a character widely
different from that of the land. At times, in peculiar states of the atmosphere, the boundary of the
horizon becomes undistinguishable, and the surface,
perfectly calm, reflects the pure light of heaven in
every part, and we seem alone in infinite space, with
nothing around that appears tangible and real save the
ship beneath our feet. At other times, particularly
in the clear waters of the tropical seas, we look downward unmeasured fathoms beneath the vessel's keel,
but still find no boundary; the sight is lost in one
uniform transparent blueness. Mailed and glittering creatures of strange forms suddenly appear, play
a moment in our sight, and with the velocity of
thought have vanished in the boundless depths. The
very birds that we see in the wide waste are mysterious; we wonder whence they come, whither they
go, how they sleep, homeless, and shelterless as they
seem to be. The breeze, so fickle in its visitings,
rises and dies away; "but thou knowest not whence
it cometh and whither it goeth;" the night-wind
moaning by, soothes the watchful helmsman with
gentle sounds that remind him of the voices of be-
loved ones far away; or the tempest shrieking and
groaning among the cordage turns him pale with the
idea of agony and death. But God is there; lonely
though the mariner feel, and isolated in his separa-

tion from home and friends, GOD is with him, often unrecognized and forgotten, but surrounding him with mercy, protecting him and guiding him, and willing to cheer him by the visitations of His grace, and the assurance of His love. " If I take the wings of the morning and dwell in the utmost parts of the sea; even there shall Thy hand lead me, and Thy right hand shall hold me."

The Ocean is the highway of commerce. God seems wisely and graciously to have ordained, that man should not be independent, but under perpetual obligation to his fellow-man; and that distant countries should ever maintain a mutually-beneficial dependence on each other. He might with ease have made every land to produce every necessary and comfort of life in ample supply for its own population; in which case, considering the fallen nature of man, it is probable the only intercourse between foreign nations would have been that of mutual aggression and bloodshed. But he has ordered otherwise; and the result has been, generally, that happy interchange of benefits which constitutes commerce. It is lamentably true, that the evil passions of men have often perverted the facilities of communication for purposes of destruction; yet the sober verdict of mankind has for the most part been, that the substantial blessings of friendly commerce are preferable to the glare of martial glory. But the transport of goods of considerable bulk and weight, or of such as are of a very perishable nature, would be so difficult by land, as very materially to increase their cost; while land communication between coun-

tries many thousand miles apart would be attended
with difficulties so great as to be practically insurmountable.   Add to this the natural barriers presented by lofty mountain ranges and impassable
rivers, as well as the dangers arising from ferocious
animals and from hostile nations, and we shall see
that with the existing power and skill of man, commerce in such a condition would be almost unknown,
and man would be little removed from a state of bar-
barism.   The Ocean, however, spreading itself over
three-fourths of the globe, and penetrating with innumerable sinuosities into the land, so as to bring,
with the aid of the great rivers, the facilities of navigation comparatively near to every country, affords
a means of transport unrivalled for safety, speed, and
convenience.   In very early ages men availed themselves of naval communication.   We find repeated
mention made of ships by Moses;* and in the
dying address of the patriarch Jacob to his sons, he
speaks of "a haven for ships;"† while Job, who
was probably contemporary with Abraham, alludes
to them as an emblem of swiftness,‡ which would
seem to imply that navigation had then attained
considerable perfection, nearly four thousand years
ago.   In profane history the earliest mention of
navigation is that of the voyage of the ship Argo
into the Euxine, which took place probably about
three thousand years ago.   What a contrast between her timorous and creeping course, and the
arrowy speed and precision of a modern Atlantic

* Numb. xxiv. 24; and Deut. xxviii. 68.   † Gen. xlix. 13.
‡ Job ix. 26.

steam-ship, rushing to her destination without asking aid from wind or tide!

The proportion which the sea bears to the land in extent of surface has been ascertained with tolerable accuracy, by carefully cutting out the one from the other, as represented on the gores of a large terrestrial globe, and weighing the two portions of paper separately in a very delicate balance. The ratio of the water to the land is found to be about $2\frac{3}{4}$ to 1: the surface of the former being about one hundred and forty-four millions of square miles, and that of the latter about fifty-two millions. Vast, however, as is the sea, and mighty in its rage, it is restrained by the hand of Him that made it. Water was once the instrument of vengeance upon a guilty world, but he hath made a covenant with man, that never again shall the waters become a flood to destroy the earth. He "shut up the sea with doors, when it brake forth as if it had issued out of the womb; when He made the cloud the garment thereof, and thick darkness a swaddling-band for it; and brake up for it His decreed place, and set bars and doors, and said, Hitherto shalt thou come, but no further, and here shall thy proud waves be stayed!"* Slight changes are, it is true, going on in the course of ages, in the relative positions of the land and sea, but these are minute in their extent and slow in their operation. By the sand and mud, which are continually brought down by the rivers and deposited in the sea, banks and points of land are formed and perpetually in-

* Job xxxviii. 8–11.

creased, as is particularly the case at the mouths of the Ganges and Mississippi; while from the same cause the bottoms of inland seas being gradually raised, the water rises in the same proportion and encroaches on the land. The port of Ravenna, once a rendezvous for the Roman fleets, has been filled up by the deposition of the Montone, a small river, so that now it is four miles from the sea. On the other hand the palace of the Emperor Tiberius at Capræa, on the opposite shore of Italy, is now wholly covered by the water: nor are our own coasts, and especially those of Holland, deficient in examples of once fertile fields, which are now rolled over by the tide.

Much ignorance prevails respecting the depth of the Ocean: in many places no length of sounding line has yet been able to reach the bottom, and, therefore, our conclusions must be formed from inference or indirect evidence. Generally, where a coast is flat and low, the water is shallow for a considerable distance, slowly deepening; on the other hand, a high and mountainous coast usually is washed by deep water, and a ship may lie almost close to the rocks. From these circumstances, as well as from the various depths actually observed by sounding, it is probable that the average depth of the sea is not greater than the height of the land, in proportion to its extent. If we were to place a thick coating of wax over the bottom of a dish, taking care to make a very irregular surface, with cavities and prominences of all forms and sizes, we should probably have a fair idea of the solid surface of the globe. Let us then pour water upon it until the surface of the water should equal that part which is exposed, and it is clear the average depth of the one would be equal to the average height of the other. But if we increase the quantity of water until the proportion is as three to one, it is evident the depth will have increased in the same ratio. We may, therefore, with high probability, conclude that, as the greatest height of the land is about five miles, the greatest depth of the water does not much exceed twelve or thirteen; while the average depth may be about two or three.

Every one is aware of the saltness of the sea. It has been assumed that its object is to prevent stagnation and putrescence. But this reason does not appear to be the correct one, for large masses of fresh water, such as inland lakes, do not stagnate. Strictly speaking, however, water cannot putrefy; when a small body of it becomes offensive, it is on account of the decomposition of vegetable or animal matters contained in it. But organized substances will decompose, and consequently become offensive in salt water as well as in fresh, as may be easily proved by experiment. Perhaps the reason for the Ocean's saltness may be the increase of its weight without the increase of its bulk; for the decrease of specific gravity of so large a portion of the globe would materially affect the motions of the earth, and perhaps derange the whole constitution of things. The increase of its specific gravity makes it more buoyant, and every one is aware with how much less effort a bather swims in

the sea than in a river. Now, superior buoyancy
seems an important advantage in a fluid which bears
on its bosom the commerce of the world. It is
highly probable, then, that our gracious God had
the convenience and benefit of man in view when
he ordained the sea to be salt. The Ocean contains
three parts in every hundred of saline matter, chiefly
muriate of soda, or the common salt of the table,
which is a chemical compound of muriatic acid and
soda. The proportion is rather large in the vicinity
of the equator. If we considered only the immense
amount of evaporation which is daily going on from
the sea, we might suppose that, like a vessel of the
fluid exposed to the sun, it would diminish in
volume and increase in saltness, until at length
nothing would be left but a dry crust of salt upon
the bottom; on the other hand, looking alone at
the many millions of tons of fresh water which
are every moment poured into its bosom from the
rivers of the earth, we might apprehend a speedy
overflow, and a second destruction by a flood. But
these two are exactly balanced; the water taken up
by evaporation is with scrupulous exactness restored
again, either directly, in rain which falls into the sea,
or circuitously, in the rain and snow, which falling
on the land, feed the mountains, streams and rivers,
and hurry back to their source. This interesting
circulation had been long ago observed by the wisest
of men: "All the rivers run into the sea; yet the
sea is not full; unto the place from whence the
rivers come, thither they return again."* And a

* Eccles. i. 7.

very beautiful and instructive instance it is of that unerring skill and wisdom with which the whole constitution is ordered and kept in order, by Him, who, with minute accuracy, "weigheth the mountains in scales, and the hills in a balance."*

The Ocean is never perfectly at rest: even between the tropics, in what are called *the calm latitudes,* where the impatient seaman for weeks together looks wistfully but vainly for the welcome breeze; and where he realizes the scene so graphically described in "The Rime of the Ancient Mariner:"—

> "Day after day, day after day,
>   We stuck, nor breath nor motion;
>   As idle as a painted ship,
>   Upon a painted ocean;"

even here the smooth and glittering surface is not at rest: for long, gentle undulations, which cause the taper mast-head to describe lines and angles upon the sky, are sufficiently perceptible to tantalize the mariner with the thought that the breeze which mocks his desires, is blowing freshly and gallantly elsewhere. The most remarkable of the motions observable in the sea, are the tides, periodical risings and fallings in the height of the surface, which take place twice every twenty-four hours, or nearly. It is now well ascertained that these motions are caused by the attraction of the sun and moon, but more particularly the latter, upon the particles of water, which moving freely among themselves with little force of cohesion, readily yield to

* Isa. xl. 12.

the attracting influence, and move towards it. The
time of high water in the open sea is about two
hours after the moon passes the meridian, owing
to the impetus which the waters have been receiving
not ceasing immediately; just as the hottest part of
the day is not noon, but about two hours after it;
and the hottest month of the year is not June, but
July. On the coast, however, high water is delayed
to a greater or less extent by the obstructions of
straits, mouths of rivers, harbors, &c. It appears
strange that the sea should be elevated, not only on
the side next the moon, but also on the side which
is diametrically opposite; so that it is high water at
the same moment on two opposite points of the
globe, each of which points follows, so to speak,
the moon in the daily revolution, and, consequently,
every part of the surface of the Ocean is raised twice
in each day. The singular phenomenon is thus
explained: the attraction of the moon elevates the
particles of water on the nearest side, by slightly
separating them from each other, which their imperfect cohesion readily admits; it also affects the
earth itself; but this being a solid body, the cohesion of its parts cannot be overcome, and the whole
mass is therefore moved towards the moon, while the
particles of water on the farther side remain, owing
to their freedom, nearly in the some position as before. The fact is, that the earth is drawn away from
the water on the remote side, and then the water is
drawn away from the earth on the near side. The
sun is greatly larger than the moon, but his attraction, owing to his great distance, does not affect the

tides to more than one-fourth of the moon's extent. When the power of these luminaries is exerted in the same direction, the result is a higher elevation, called the spring-tide: and for the reason already explained, the same occurs when they are in opposite quarters of the heavens. On the other hand, when they are in quadrature, that is, when apparently separated by just one-fourth of the heavens, the influence of the sun neutralizes, in the ratio of one-fourth, that of the moon; and hence we have the lowest tides, called neap-tides, soon after the first and third quarters of the moon.

Local circumstances greatly affect not only the time, but also the height of the tides. In some long bays, which grow gradually narrower, in the form of a funnel, the whole of the increased water which entered the mouth of the bay, being confined within very narrow limits, rises rapidly to a great height. Near Chepstow, in the Bristol Channel, for example, the tide rises from 45 to 60 feet, and on one occasion, after a strong westerly gale, it even reached to 70 feet. Again, in the Bay of Fundy, in North America, the spring-tides sometimes rise to the astonishing elevation of 120 feet. At the mouths of some large rivers, where the shore is very level to a considerable distance inland, the tide rolls in under the form of one vast wave, which is called the bore; something of this kind occurs in Solway Frith on our own coast; and it is said that if, when the tide is coming in, a man upon a swift horse were placed at the water's edge, and bidden to ride for his life, the utmost efforts of his steed would not preserve

him from the overwhelming wave. Through the
Pentland Frith, between Scotland and the Orkney
Islands, the spring-tide rushes at the rate of nine
miles an hour. The tide in inland seas is so slight
as to be scarcely observable, probably owing to the
smallness of the volume of water which they contain; and hence the astonishment which the soldiers
of Alexander, accustomed to the equable condition
of the Mediterranean, felt, when at the mouth of
the Indus, they beheld the sea swell to the height
of thirty feet.

That some purpose, important in the constitution
of our world, is effected by these periodical ebbings
and flowings of the mighty sea, is highly probable;
but our acquaintance with the arcana of nature is
too slight to point it out. In navigation they are
useful; the flood-tide permitting ships to sail up
rivers, even when the wind is adverse, and often
admitting deep vessels to pass into harbors, over
banks or bars, impassable at the ordinary depth of
the water.

Besides the tides, the sea has other motions of
great regularity, called currents. The principal of
these is the notable Gulf-stream, a strong and rapid
river, as I may say, in the sea, whose banks are
almost as well defined as if they were formed of
solid earth, instead of the same fickle fluid as the
torrent itself. It first becomes appreciable on the
western coast of Florida, gently flowing southward
until it reaches the Tortugas, when it bends its
course easterly, and runs along the Florida Reef,
increasing in force, till it rushes with amazing

rapidity through the confined limits of the Strait of
Florida, and pours a vast volume of tepid water into
the cold bosom of the Atlantic. Here, unrestrained,
it of course widens its bounds and slackens its speed,
though such is its impetus that it may be distinctly
perceived even as far as the Great Bank of Newfoundland. Nor is its strength then spent; for
many curious facts seem to warrant us in con-
cluding, that even to the coasts of Scotland and
Ireland, and down the shores of Western Europe,
this mighty marine river continues to roll its wonderful waters. The temperature of this current is
much higher than that of the surrounding water, and
this is so uniformly the case that an e' 'ance into
it is immediately marked by a sudden rise of the
thermometer. Another unfailing token of its presence is the Gulf-weed (*Sargassum vulgare*), which
floats in large fields, or more frequently in long
yellow strings in the direction of the wind, upon
its surface. The cause of this vast and important
current seems to be the daily rotation of the earth.
If we turn a glass of water quickly upon its axis, we
shall perceive that the glass itself revolves, but that
the particles of water remain nearly stationary, owing
to the slightness of their cohesion to the glass. To
a very minute insect attached to the vessel, it would
seem that the water was rushing round in an opposite direction while the glass remained stationary.
Now the earth is whirled round with great rapidity
from west to east, and the greatest amount of this
rapidity is of course at the equatorial regions, being
the part most remote from the axis: but the par-

ticles of water, for the same reason as those in
the glass, to a certain extent, resist the influence
of this rotation, and appear to assume a motion
in the opposite direction, from east to west. With
respect to all the phenomena to be explained, this
apparent motion is exactly the same as if it were
real, and we shall consider it so. Now, examine
a globe, or a map of the Atlantic, and you will see
that this westerly "set" of the equatorial waters,
meeting the coast of South America, is slightly
turned through the Caribbean Sea, until it strikes
the coast of Mexico, which, like an impregnable
rampart, opposes its progress. The stream, impelled
by the waves behind, must have an outlet, and the
form of the shore drives it round the northern side
of the Gulf of Mexico, until it is again bent by the
peninsula of Florida. But here the long island of
Cuba meets its southerly course, and, like the hunted
deer, headed at every turn, the whole of the broad
tide that entered the Gulf, now pent up within the
compass of a few leagues, rushes with vast impetus
through the only outlet that is open, between Florida
and the Bahamas. It is as if we propelled with
swiftness against the air a wide funnel, the mouth
being outwards, the tube of which was long and
tortuous, and which terminated at length nearly at
right angles to the mouth: it is easy to imagine
that a strong current of air would issue from the
tube, exactly as the waters of the Gulf-stream do
from their narrow gorge. The waters of the Pacific have the same westerly flow, but its force is
broken, without being turned, by the vast assem-

blage of islands which constitute the Eastern Archipelago; it may, however, be recognized in the Indian Ocean, and when bent southward by the African coast, and confined by the island of Madagascar, it forms a current of considerable force, which rounds the Cape of Good Hope, and merges into the Atlantic. Besides these, there are other more local currents, which are not so easily explained, such as that which constantly flows out of the Baltic, and that which flows into the Mediterranean. In each of these cases, while the main current occupies the middle of the channel, there is a subordinate current on each side close to the shore, which sets in the opposite direction.

As in the case of the tides, it is obvious how serviceable these motions of the sea often are in aiding navigation, particularly as they are most strong and regular in latitudes where calms often prevail.

And this leads us to consider the action of the winds upon the sea, which, though affecting only the surface, are the most powerful agents in producing the irregular motions of this element. By them the freighted bark, with her hardy crew, is wafted to the wished for haven; and by them the crested billows are roused up, which dash her upon the sharp-pointed rocks, or swallow her up in fathomless depths, leaving none to record her destiny. The origin of wind has usually been attributed to the rarefaction of the air by heat: a stratum of air near the earth being heated by the sun's rays, or by radiation from the surface, becomes lighter, and consequently rises to a

higher elevation. The empty space thus left is instantly filled by the surrounding air rushing in,
pressed by the weight of the atmosphere above: this
motion communicated to the air, has been supposed
to constitute a wind blowing in the direction of
the spot where the heat was generated. It must
be confessed, however, that the cause thus adduced
does not seem adequate to produce the effects attributed to it; though probably some of the currents of the air are owing to variations of its temperature. And as these variations are perpetually
occurring, dependent on causes which are difficult
to detect, and as the aerial currents resulting from
them act and react on each other, variously modi-
fying their direction, force, and duration, the ordinary winds are irregular and inconstant even to
a proverb. Some observations, however, recently
made, have revealed some particulars of a highly-
interesting character, concerning the winds of the
temperate zones: one of which is, that they blow
in a circular direction; that is, the course which
a storm has taken, if marked out on a map or
globe, would describe a circle, often of many degrees in diameter. The direction of the gale in
the circle is not arbitrary but seems to be invariably from north to west, south, and east, in the
northern hemisphere, and in the opposite course
in the southern. These winds appear to be intimately connected with magnetism: it is a curious
fact, that, in the midst of the southern Atlantic,
where magnetic influence is at the lowest degree
of intensity, storms are unknown, while the meri-

dians of the magnetic poles, that of the American cutting the West Indies, and that of the Siberian the China Sea, are peculiarly liable to tempests; the hurricanes of the former, and the typhoons of the latter, being well known.* It is pretty certain, also, that the changes in the atmosphere produced by electricity, which is but another development of the same principle as magnetism, have considerable influence in the production of the variable winds of temperate regions. Our knowledge of these sub-jects, however, is yet in its infancy; and though in all ages until the present, navigation has been entirely dependent on the aid of the winds, no laws for their certain prognostication have yet been dis-covered, and much obscurity, at least in detail, still hangs over their production. But within the tro-pical regions there are winds which possess great regularity, and may be depended upon with nearly the same precision as the great marine currents already noticed, which indeed they very closely re-semble, not only in their direction and their utility, but also in their origin. I refer particularly to the Trade-winds, so named from the facility they afford to commerce, which blow constantly, within the tro-pics, from the north-east on the north side of the equator, and from the south-east on the south side, the two currents merging near the line into one, which takes an easterly direction. The dividing line, how-ever, is not exactly at the equator, but a little to the north of it. The air in the equatorial regions be-comes strongly heated by the rays of the vertical sun,

* See Reid on Storms.

and rises; while that from the polar regions moves in to supply its place: thus a nothern and southern current are produced towards the equinoctial. But the earth is revolving from west to east, and the equatorial parts are, as we have before seen, those in which the velocity is greatest: the free air cannot at once acquire this velocity, and is left behind; the effect being that an apparent motion in the contrary direction is given to it, which, combining with the one already possessed by the polar currents, makes the direction of the northern one north-east, and of the southern south-east. The point directly beneath the sun, also, is continually travelling westward, which increases the effect. The heat radiated from the surface of large masses of land being superior to that from the sea, while the former is subject to much variation from differences of elevation, and other circumstances, the trade-winds are disturbed, and become very irregular in the vicinity of land; but in open sea they blow with much precision.

A singular deviation from the uniformity of the trade-winds occurs in the Indian Ocean, which it seems difficult to explain. From 30° south latitude, to within about 10° of the equator, the trade is pretty constant from the south-east; but to the north of the latter parallel, the wind blows six months from the north-east, namely, from October to April, while, during the remainder of the year, from April to October, it blows with equal pertinacity in a direction diametrically opposite. These are called respectively the north-east and

south-west *monsoons;* but the former is the regular
trade—the latter alone is the anomaly, and needs
explanation. The cause usually assigned is, the
rarefaction of the air on the continent of Asia
during the summer months, when the sun is north
of the equator; the air from the Indian · Ocean
flowing in to supply its place. This would sufficiently explain why the wind should be southerly,
but leaves its westerly inclination entirely unac-
counted for; and this seems the more inexplicable,
because one would suppose that the air over the
burning deserts of Arabia and North Africa would
be much more heated, and that the direction of the
supplying current would be south-east. Strange,
however, as the fact is, it is perfectly uniform in
its occurrence, and is obviously a very gracious
ordination of God's beneficent providence, in diminishing the uncertainties of navigation.

There is yet another phenomenon connected with
the wind, in the climates of which we speak, that
requires notice; it is the alternation of the land
and sea-breezes. Every one who has resided near
the coast in tropical countries is aware of the eagerness with which the setting in of the sea-breeze is
looked for. Usually about the hour of ten in the
forenoon, when the heat of the sun begins to be
oppressive, a breeze from the sea springs up, in-
vigorating and refreshing the body by its delightful coolness, and continues to blow through the
whole day, gradually dying away as the sun sinks
to the horizon. Then, about eight in the evening,
an air blows off the land until near sunrise; but this

is somewhat variable and irregular, always fainter than the sea-breeze, and dependent on the proximity of mountains. The application of what has been already said of the causes of wind in general will readily be made to these particular cases, the air on the surface of the water being cooler during the day, and that on the mountains during the night. Either is a grateful alleviation of the oppressive sultriness of the climate.

But for the winds, the surface of the sea would ever present, notwithstanding its intestine motions, an unbroken and glassy smoothness. The playful ripple which breaks the moon's ray into a thousand sparkling diamonds, and the huge billows that rear their curling and cresting summits to the sky, would be alike unknown. If the direction of the breeze were exactly horizontal, it is difficult to imagine how the surface could be ruffled at all; but doubt-less the wind exerts an irregular *pressure* obliquely upon the water, a few particles of which are thus forced out of their level above the surrounding ones: these afford a surface, however slight, on which the air can act directly, and the effect now goes on in-creasing every moment, until, if the wind be of suf-ficient velocity, the mightiest waves are produced.*

* The perpendicular elevation of even the highest waves is, however, much overrated. Viewed from the deck of a vessel, the immense undu-lating surface causes them to appear much higher than they are; while the ever-changing inclination of the vessel itself produces a deception of the senses, which increases the exaggeration. Experienced practical men have, however, made some observations, which show us their height. Taking their station in the shrouds, they have proceeded higher and higher, until the summit of the loftiest billow no longer intercepted the

" For he commandeth, and raiseth the stormy wind, which lifteth up the waves thereof. They [the mariners] mount up to the heaven, they go down again to the depths: their soul is melted because of trouble. They reel to and fro, and stagger like a drunken man, and are at their wits' end." The Holy Spirit thus alludes to the terrific raging of the tempest as eminently calculated to draw man's attention to the power and majesty of God, while the wondrous deliverances He has so often wrought from its fury, are so many claims on man's grateful love and praise.

Let us, then, in contemplating a few of the innumerable objects of interest which the ocean presents to us, endeavour in dependence on His own gracious aid, to recognise His hand, to discern the greatness of His power in creating and upholding all things; His unerring skill and wisdom in arranging and carrying out His designs; and the careful and provident benevolence which He continually exercises towards the sentient part of His creation. The varied tribes of living beings that throng the deep, from the wallowing whale to the luminous animalcule, visible but as a sparkling point; the multifarious forms of marine vegetation, displaying

view of the horizon. After watching for a sufficient length of time to verify the deductions, they descended, and measured the height of the point of sight from the ship's water-line; deducting half of this distance for the depression of the hollow below the level of the surface, the remainder gives the elevation of the highest wave. It is thus found that waves do not usually exceed six feet in height, except when cross-waves over-run each other; and probably in no case do the very loftiest rise above ten feet above the general level.

3

exquisite structure and elaborate contrivance; the golden sands of the smooth shore, the hoary cliffs hollowed into caverns by the restless billows, and not least, the restless billows themselves, speak to us, in language not to be mistaken, of the glorious attributes of the Mighty God, "the Lord of Hosts, which is wonderful in counse⁻ and excellent in working."

THE SHORES OF BRITAIN.

BEFORE we launch forth to investigate the wonders of the vast Ocean, a little time will not be misspent in observing a few of the curious productions which it brings to our very doors. We shall greatly err, if we suppose that only in distant parts of the world the works of God can be so studied as to illustrate His infinite power, and skill, and benevolence: we may have to search distant regions to find the giants of the deep, the huge whale, the Indian cuttle, or the island madrepore; but in the most minute crustacean that hops above the retiring wave, or the most fragile shell that lies upon the shingle, there is the indelible impress of the mind and hand of God. Indeed, it may be asserted, that of two created objects of different magnitude, but possessing similar organs, equally adapted to their requirements, that one in which these organs are of minute size is the more calculated to excite our admiration. Our own shores swarm with little creatures of many kinds, some so small as to escape the eye of any one but a naturalist, which yet are well worthy of being examined and studied. Take one example. Walking along a sea-beach, where the loose shingle rattles under the retiring waves, we may find a

minute beetle, known to entomologists by the name
of *Aepus fulvescens*, whose habits may well excite
our astonishment. Formed like all other beetles,
to breathe air alone, it deserts the haunts of its fellows, and betakes itself to the sea, choosing to dwell
among the pebbles so low down on the beach that
the water covers it constantly, except for a day or
two twice every month, when, at the lowest ebb of
the spring-tide, it is for a few minutes exposed.
Now, during the weeks of its submersion, how does
this little creature breathe? Oxygen it must have,
or it will assuredly die. Many of the beetles that
shoot hither and thither in our fresh-water ponds
are clothed with a coat of thick but very fine down,
in which air is entangled and carried beneath the
surface. But our little *Aepus* is not furnished with
a coating of down. If we examine it, however,
with a magnifier, we shall discover that its whole
body and limbs are studded with long, slender
hairs, and when it plunges under water, each of
these hairs carries with it a little globule of air
from the atmosphere, and these, uniting, form a
bubble of air surrounding the body of the insect,
and serving it for respiration. But, subjected to
the rolling of the tide, it would be liable to be
perpetually washed away from its dwelling-place,
were there not an especial provision graciously made
for its stability. For this end the feet are furnished with claws of unusual size, to cling firmly
to the projections of the stones, and in addition
to these the last joint but one of the feet has a
long curved spine meeting the claws, giving it an

extraordinary power in grasping, as well as aiding it
in obtaining its prey. In other respects, with regard
to its eyes, its antennæ, its jaws, we shall find, if
we carefully examine it, that, minute as it is, being
scarcely an eighth of an inch long, its wants have
been accurately remembered and well supplied. A
few other British insects, likewise very small, display similar instincts, some of them inhabiting holes
in the sand, very near low-water mark, and therefore entirely submerged a great portion of their
time.

On our rocky shores may be found in abundance
creatures still more minute than these, whose manners, lively and sportive, are highly interesting. I
allude to the marine *Entomostraca*, or insects with
shells, and particularly to those of the genus *Cythere*,
scarcely any of which exceed in diameter a large
pin's head, and most of them are not equal to that
of a small one. Imagine a pair of bivalve shells of
this size, irregularly oval, or kidney-shaped, from
which, slightly separated, protrude four pairs of little
curved claws, or feet, most delicately fringed, and
kept in constant motion; and from one end a pair
of jointed antennæ. Mr. Baird, who has attentively
studied their manners, gives the following pleasing
account of them:—"These insects are only to be
found in sea-water, and may be met with in all the
little pools amongst the rocks on the sea-shores.
They live amongst the *Fuci* and *Confervæ*, &c., which
are to be found in such pools; and the naturalist
may especially find them in abundance in those
beautiful clear little round wells which are so often

D

to be met with, hollowed out of the rocks on the
shores of our country, which are within reach of
the tide, and the water of which is kept sweet and

MARINE ENTOMOSTRACA (*Cythere albo-maculata* and *Cyclops chelifer*).

wholesome by being thus changed twice during
every twenty-four hours. In such delightful little
ponds, clear as crystal when left undisturbed by the
receding tide, these interesting little creatures may
be found often in great numbers, sporting about
amongst the confervæ and corallines which so
elegantly and fancifully fringe their edges and de-
corate their sides, and which form such a glorious
subaqueous forest for myriads of living creatures
to disport themselves in. Sheltered amongst the
"umbrageous multitude" of stems and branches,
and nestling in security in their forest glades, they
are safe from the fury of the advancing tide, though
lashed up to thunder by the opposing rocks which
for a moment check its advance; and weak and
powerless though such pigmies seem to be, they
are yet found as numerous and active in their
little wells, after the shores have been desolated

by the mighty force of the tide which has been driven in, in thunder, by the power of a fierce tempest, as when the waves have rolled gently and calmly to the shore in their sweetest murmurs. These insects have never been seen to swim, invariably walking among the branches or leaves of the *confervæ* or *fuci*, amongst which they delight to dwell; and when shook out from their hiding-places into a bottle or tumbler of water, they may be seen to fall in gyrations to the bottom, without ever attempting to dart through the watery element, as in the case with the *Cyprides*. Upon reaching the bottom they open their shells, and creep along the surface of the glass; but when touched or shaken, they immediately again withdraw themselves within their shell, and remain motionless."* The *Cyprides*, here alluded to in comparison, are species very closely resembling these, inhabiting abundantly every stagnant ditch and pool of fresh water. They have their antennæ and feet beautifully feathered with long fringed bristles, by aid of which they swim with much vivacity. In exactly similar situations to those above described are found other *Entomostraca*, marine species of the genus *Cyclops*, almost equally minute, and equally interesting. Like their kindred of the same genus found in fresh waters, and which are so numerous in the water conveyed into London that we swallow them daily, these swim with ease, progressing by sudden bounds made with great vigour and effect. Mr. Baird notices of one marine species (*C. depresses*), which he

* Mag. Zool. and Bot. ii. 141.

found in Berwick Bay, that its motion is peculiar.
"It generally swims on its back, and instead of
darting *forward* through the water, as the other
species of *Cyclops* do, it springs with a bound from
the bottom of the vessel, where it rests when undisturbed, *up to the surface* of the water. For this
purpose it curls its body up into the form of a ball,
and then, suddenly returning to the straight position, springs with a sudden bound from the bottom
to the surface, falling gradually down again to the
same place from which it sprung." It is a remarkable character of all these pretty little *water-fleas*,
that they have but a single eye, which is generally
of a bright crimson hue, sparkling like a little ruby,
and is set in the front of the head. Any of my
inland readers, who may have no opportunity for
sea-side researches, may form a very good idea of
the form and habits of these agile "minims of existence" by pulling up a handful of the common duck-
weed from a stagnant pool, and putting a pinch of
it into a clear glass phial, nearly filled with water:
numbers of the fresh-water *Entomostraca* will be
almost certain to swim out; and the sight will amply
repay the trouble of procuring them, especially if
viewed with a microscope, or even a common magnifying glass.

Probably the objects which would first arrest the
observation of one who for the first time visited
a rocky shore, would be, after the broad element
itself, the marine plants which in such abundance
and variety clothe the submerged rock. At a glance
we perceive that they are singular productions; the

vast size of some, the strange and uncouth forms
of others, and the extreme delicacy and vivid hues
of many, cannot fail to attract attention: and it
needs not the additional knowledge that many of
them are pressed into the service of man to assure
us that they are not less worthy of the consideration
of rational beings than others of the glorious works
of God. "Viewing these tribes," observes Dr. Greville, "in the most careless way, as a system of subaqueous vegetation, or even in a merely picturesque
light, we see the depths of ocean shadowed with
submarine groves, often of vast extent, intermixed
with meadows, as it were, of the most lively hues;
while the trunks of the larger species, like the great
trees of the tropics, are loaded with innumerable
minute kinds, as fine as silk, or transparent as a membrane."* In stating some particulars of the history
of but a few of the species found on our own shores,
I hope to show that the contempt which has been,
even to a proverb, cast upon the "vile sea-weed,"
is very much misplaced. It is only a contracted
mind, governed by debasing selfishness, which mea-
sures the esteem in which it holds any object by
the degree to which it ministers to the comfort or
profit of man; the instructed Christian will feel a
higher gratification in the thought that the perfec-
tions of God shine forth more luminously the more
His handiwork is examined. It was no selfishness
that prompted the Sons of God, when they saw this
beautiful and glorious world, fresh in its unsullied
prime, come from the hands of its Maker,—to sing

* Algæ Britannicæ. Intr.

D 2

together, and all the morning stars to shout for joy.
Yet we may, with adoring gratitude, recognise the
love which remembers man, and provides many natural objects for his appropriation; endowing them
with qualities which his intelligence discovers to be
useful, and which alleviate the privation and toil of
his fallen condition.

A substance called kelp, an impure carbonate of
soda, important in the manufacture of soap and of
glass, is the produce of these "worthless" weeds.
Some years ago, the coasts and islands of Scotland
yielded 20,000 tons of this valuable substance annually, which was worth ten pounds sterling per
ton; but through the increased consumption of *ba-
rilla*, an alkali imported from Spain, it has somewhat diminished. The autumnal storms detach large
quantities of *Algæ* (a general name applied to all
the sea-weeds), which are washed ashore. The
inhabitants of the coast, aware of their value,
hurry down to secure the riches thus freely presented, and either cast them on their fields as a valuable manure, or burn them into kelp. In Scotland, the kelp-kiln is nothing but a round pit, dug
in the sand or earth on the beach, and surrounded
by a few loose stones. In the morning a fire is
kindled in this pit, generally with the aid of turf
or peat. The fire is gradually fed with sea-weed,
in such a state of dryness that it will merely burn.
In the course of the day, the furnace becomes
nearly full of melted matter, and iron rakes are
then drawn rapidly backward and forward through
the mass to compact it, or bring the whole into an

equal state of fusion. It is then allowed to cool, and having been taken out and broken to pieces, it is carried to the storehouse to be shipped for market. The general yield of this alkali is one-fifth of the weight of the ashes from weeds promiscuously collected; but from one species, the Sea-wrack, or Black-tang (*Fucus vesiculosus*), one of the most abundant on our coast, the ashes yield half their weight of alkali. The Sea-wrack is of a dark-green hue, bearing long, flat, and narrow fronds, resembling leaves, divided into branches, and having a midrib running through the centre; the leaf-like branches terminate in large yellow oval receptacles, containing many seeds, enveloped in a thick mucus. But its chief peculiarity is, that the substance of the frond swells at irregular intervals into oval air-cells, always arranged in pairs, one on each side of the midrib. The Dutch use this sort, and another called Black-wrack (*F. serratus*), to pack their lobsters; the latter, however, is preferred, on account of its containing less mucus, and therefore being less liable to fermentation.

Scarcely inferior in its alkaline properties to the Sea-wrack is the Knotted-wrack (*F. nodosus*). The fronds look like slender stems, swelling at intervals into oval bulbs or air-vessels. Boys amuse themselves occasionally by cutting off these nodules in a diagonal direction, to make them into whistles. They are too tough to be burst by the pressure of the fingers, like those of the Sea-wrack; but if stamped on, or put into the fire, they explode with a loud report. The seed-vessels are large, oval, and yellow, resembling those of the last, placed on foot-stalks.

One of the most common species of our coasts is the long, string-like Sea-lace, or, as the Orkney people call it, Sea-catgut (*Chorda-filum*). It usually grows in water some fathoms deep, attached to stones at the bottom, yet reaching to the surface: indeed, it sometimes attains the length of forty feet; and this is believed to be the growth of a single summer, as it is an annual plant. Its structure is highly curious; at first sight it appears a simple cylindrical tube, of an olive colour, about as thick as whipcord, but occasionally thicker: on examination, however, this hollow stem is found to be composed of a flat thin ribbon, abouth one-sixth of an inch in width, spirally twisted into a tube, the edges exactly meeting each other, and adhering with sufficient firmness to allow of the whole stem being skinned without separating: in this state it is twisted and dried, when it possesses a strength and toughness that adapt it for fishing-lines. In Norway it is collected as food for the cattle. The upper portion usually floats on the surface, or rather immediately beneath it, often in such abundance as to form large meadows, as it were, which obstruct the progress of boats. The fructification of this species long defied the investigations of botanists; but it is now ascertained to consist of little pear-shaped capsules, imbedded in the surface, and much crowded, which the gradual melting away of the skin allows to escape. One of the most interesting

circumstances connected with the history of the sea-
plants is, the beautiful and varied apparatus with
which many of them are provided for securing
buoyancy. It seems to be essential to their health
that they should at least approach the surface, but
as their substance is specifically heavier than water,
many of them are greatly lengthened, and furnished with hollow vessels inflated with air, by
which their weight is diminished. These differ
much in form and position in the various tribes;
in the Sea-wrack (*F. vesiculosus*), we saw them take
the form of bladders, arranged in pairs on each side
of the midrib; in the Knotted-wrack (*F. nodosus*) ·
the stem swells at intervals into hollow bulb-like
dilatations, while in the long Sea-lace before us.
the same end is answered by dividing the hollow
tube into chambers, interrupted at short distances
by portions of the solid substance of the frond;
the cavities being filled in some unknown manner
with air, probably hydrogen generated by the plant
itself.

Many of the *Algæ* are rather extensively used as
food; and though to one unused to such diet they
would in general seem to offer little temptation to
the appetite, the poorer natives, not only of our own
but of other shores, eat them with much relish. Let
us not despise their taste, though differing from our
own, but rather adore the beneficence of God, who
has supplied in much abundance an additional source
of nutriment, and has conferred on the recipients
of His bounty the taste requisite for its enjoyment.
From the quantity of saccharine matter which they

contain, many of these plants are highly nutritive, and cattle often feed on them with greediness. One of the species most extensively eaten is that known in Scotland by the name of Dulse (*Rhodomenia palmata*). It exhibits the appearance of a very thin, membranaceous leaf, irregularly oblong, of a purplish colour, or sometimes rosy-red: there is no rib, but the substance is uniform; it grows from three inches to a foot in length. Before the introduction of tobacco, this leaf was rolled up and chewed in the same manner as the Virginian leaf is at present. It is an important plant to the inhabitants of Iceland; they wash it thoroughly in fresh water, and dry it in the air, when it becomes covered with a white powdery substance, which is sweet and palatable; it is then packed in close casks, and preserved for eating. It is used in this state with fish and butter, or else, by the higher classes, boiled in milk, with the addition of rye-flour. In Kamschatka, a fermented liquor is produced from it. It is extremely common on all our coasts, and being frequently washed on shore, is sought with avidity by the cattle: sheep sometimes go so far in the pursuit of it at low water as to be drowned by the returning tide. This species, with another which I am about to describe, was, until recently, so much esteemed by our northern countrymen, that it was publicly sold in the cities as an article of regular consumption. The cry of "Buy dulse and tangle," resounded at no very distant period even through the streets of Edinburgh. The latter is the sea-weed, usually called in England the Sea-girdle, and in the

THE SEA-GIRDLE (*Laminaria digitata*).

Orkneys Red-ware (*Laminaria digitata*). It is very common, growing chiefly in deep water, where it is protected from the heavy action of the waves. Its appearance is singular: from a number of little rootlets, which grasp with great tenacity the naked rock, springs a straight olive-brown stem, sometimes as thick as a man's wrist, and three or four feet long: at the summit it dilates into a broad cartilaginous leaf, oblong in form, and palmated, or

divided into numerous irregular strips; it is endowed
with the power of renewing its frond if the latter
be accidentally destroyed. Mr. Johns observes,*
that of all the various kinds of sea-weeds thrown
on shore during a storm, Tangles are the most abundant: a fact which he explains by the ravages of
a species of limpet (*Patella lævis*) upon their stems
and rootlets. When cooked, the young stalks are
said to be not unpleasant, and they are boiled and
given to cattle. But, as we are informed by Mr.
Neill, "in Scotland the stems are sometimes put to
rather an unexpected use, the making of knife-
handles. A pretty thick stem is selected, and cut
into pieces about four inches long. Into these, while
fresh, are stuck blades of knives, such as gardeners
use for pruning and grafting. As the stem dries, it
contracts and hardens, closely and firmly embracing
the hilt of the blade. In the course of some months
the handles become quite firm, and very hard and
shrivelled, so that when tipped with metal they are
hardly to be distinguished from hartshorn."

Much resembling this species, but immensely
larger, is the plant which has received the name
of Sea-furbelows (*L. bulbosa*). A single specimen,
fresh from the sea, is a heavy load for a man's
shoulder: and one which was measured by Mrs.
Griffiths, when spread out, covered a circular space
of twelve feet in diameter. The great weight of the
frond in this species requires extraordinary support
against the force of the waves, which else, having
so strong a purchase, would soon overturn it. To

* Botanical Rambles, p. 286.

guard against this, the ordinary mode of attachment
to the rock would be insufficient; and, instead of
the primary root, the base of the stem is swollen out
into a large hollow bulb, the extended surface of
which putting forth powerful rootlets from every

THE SEA-FURBELOWS (*Laminaria bulbosa*).

part enables the plant to defy the violence of the
winter storm. It is a fact worthy of our notice and
admiration, that nothing of the kind takes place
while the plant is young and small; it is only when
it acquires size and weight, or, in other words, it is
only when additional support becomes needful, that
this extraordinary but most effective contrivance is
resorted to. The English name of the species is

4   E

derived from the edge of the stem, which is greatly
dilated and curled into tortuous waves or plaits.

A long, narrow, ribbon-like leaf, with a thick midrib, grows on the coast of Scotland, where it is called
Hen-ware, as well as on the northern shores of Ireland, where it receives the appellation of Murlins.
It is the *Alaria esculenta* of botanists. It is of a
transparent yellow-green, and in the herbarium dries
without any change, and has a very beautiful appearance. The midrib is the part usually selected
for eating, but Mr. Johns gives us a somewhat unfavourable notion of its quality. "While walking,"
he observes, "round the coast near the Giants'
Causeway, I once observed a number of men and
women busily employed near the water's edge; and
on inquiring of my guide, found that they were
providing themselves with food for their next meal.
Being curious to discover what kind of fare the
rocks afforded, I stopped one of the men, who was
going home with his bundle, and asked him to give
me a bit to taste, prepared in the way in which it
was generally eaten. He accordingly stripped off all
the expanded part of a long and narrow leaf, and
presented me with a stem, or midrib. It was, I
must confess, as good as I expected; but at best a
very sorry substitute for a raw carrot, combining
with the hardness of the latter the fishy and coppery
flavour of an oyster. I made a very slight repast, as
you may suppose; and, after having given the man a
few pence for his civility, continued my walk. My
guide, however, seemed to think, that if I did not
choose to enjoy to the full the advantage which I had

purchased, there was no reason why he should not. He accordingly stayed behind for a minute or two, and when he rejoined me, was loaded with a supply of the same plant, which he continued to munch with much apparent relish as we pursued our walk."* Mr. Drummond, however, it must not be concealed, gives a somewhat different account, both of the part which is eaten and its flavour, and as his observations refer to the coast of Antrim, it is not easy to account for the conflicing statements, except by supposing some variation of taste in different neighbourhoods or individuals. The latter gentleman says, "It is often gathered for eating, but the part used is the leaflets, and not the midrib, as is commonly stated. These have a very pleasant taste and flavour, but soon cover the roof of the mouth with a tenacious greenish crust, which causes a sensation somewhat like that of the fat of a heart or kidney. These leaflets are quite membranaceous when young, but in full-grown plants are fleshy, and at their middle a quarter of an inch or more in thickness."†

The Dulse of the Scottish coast, which was just now described, must not be confounded with the Dulse of the southern shores of England. This is a very different plant (*Iridæa edulis*), having little resemblance to it, except in being eatable. It consists of a short stem expanding into an oval leaf, without rib or veins, sometimes a foot and a half long, and eight or ten inches wide. It is thick and fleshy, of a deep blood-red hue, the surface smooth and glossy. It is not frequently found, however, in

a perfect state, the specimens being generally torn
and perforated in every possible way. These defects
have usually been attributed to the munching of
crabs, which are said to be fond of it; but Mr.
Drummond is of opinion that portions spontaneously
separate from the frond and drop out. Like many
other *Algæ*, it diffuses, when moist, a strong smell
of violets. The fishermen pinch the fleshy frond
between heated irons, and eat it; its taste is said to
resemble that of roasted oysters. Its deep colour
may yet be found useful in the arts: Mr. Stackhouse observes,* "The most surprising quality of
this plant, and one that will probably render it of
service in dyeing, I discovered by accident. Having
placed some of the leaves to macerate in sea-water,
in order to procure seeds from it, I perceived, on
the second day, a faint ruby tint, very different from
the colour of the plant, which is a dull red, inclining
to chocolate colour. Being surprised at this, I continued the maceration, and the tint grew more vivid,
till at last it equalled the strongest infusion of cochi-
neal. This liquor was mucilaginous, and had a remarkable property of being of a changeable colour;
as it appeared a bright ruby when held to the light,
and a muddy saffron when viewed in a contrary direction: this probably arose from a mixture of the frond
in the liquor. I endeavoured to ascertain its dyeing
powers by the usual process without success; as the
quanity of tinging matter was not sufficient; though
if attempted at large, and properly evaporated, it

* Nereis Brit. p. 58, as quoted by Turner, Hist. Fucorum, ii. 113;
but I could not find the observation in Stackhouse.

might be made sufficiently strong. However, an ingenious chemical friend (the Rev. W. Gregor) assures me he has procured a fine lake from an infusion of it by means of alum."

One or two species of the genus *Porphyra* are brought to our tables, stewed under the name of Laver, and are thought a delicacy. Mr. Drummond informs us that *P. laciniata*, called Sloke in Ireland, is gathered during the winter months only, the fronds being too tough in the summer. After being properly cleaned, it is stewed with a little butter, to prevent its getting a burnt flavour, and is brought to Belfast, where it is sold by measure usually at the rate of fivepence per quart. Before being brought to table, it is again heated with an additional quantity of butter, and is usually eaten with vinegar and pepper. *P. vulgaris* is worthy of notice on account of the extreme difficulty with which it is preserved in a herbarium in a complete state: "not that there is any difficulty in spreading and going through the other steps of the process, but because when it has nearly arrived at the last stage of drying, a moment's exposure to the air will cause it to contract so instantaneously, that the edges of the paper are immediately drawn towards each other; and if attempted to be restored without the whole being first damped, the specimen tears through the middle, and becomes of little value. The edges of the plant adhere strongly to the paper when dry, or nearly so; but the centre does not adhere at all, and being as fine as gold-beater's leaf, though having considerable strength, it at once loses the little moisture it possesses, on

coming in contact with the air, and contracts with
a force remarkable when we consider its extreme
thinness. If the paper be thin, its four corners will
in a moment be brought almost in contact with each
other." The best method of obviating this inconvenience is said to be, when we suppose it is almost
dry, to have a flat book held open, and the pressure
being taken off, to remove the specimen along with
the drying-paper covering it, as quickly as possible
into the book, which must be instantly shut, and not
opened till the next day, or till we know that it is
thoroughly dry.*

There is a substance which has been lately intro-
duced as an article of commerce, intended as a substitute for Iceland moss, and sold by the London
druggists by the name of Carrageen moss; notwithstanding its name, however, it is a true *Alga, Chondrus crispus.* It is an exceedingly variable species,
but its most usual form is that of a flat leaf, spreading
somewhat triangularly, or rather so as to give to its
outline the figure of one-fourth of a circle: the edge
is branched into numerous flat segments overlapping
one another. When viewed under water, in a growing state, it gives out beautiful prismatic hues. Containing a large quantity of gelatine, it has been suc-
cessfully applied, instead of isinglass, in the making
of blanc-mange and jellies. A fucus, probably allied
to this, found at the Cape of Good Hope, is boiled
into a jelly, and, being mixed with sugar and the juice
of lemons or oranges, makes a very agreeable dish.

I shall notice a few other *Algae,* remarkable either

* Drummond.

for singularity or beauty, and then dismiss these interesting tribes. The common Sea-thong (*Himanthalia lorea*), so generally distributed, is worthy of
observation on account of its curious mode of growth.
From a shallow cup, affixed to the rock by a short
foot-stalk, spring two or three long, olive-coloured
straps, each of which becomes divided into two, and
each of these into two more, in succession: these
attain commonly the length of eight or ten feet,
and have been asserted to reach even twenty feet.
The thongs have been usually considered the fronds
of this species; but Dr. Greville thinks that the singular cup is the true frond, and the thongs the receptacles of the seed greatly lengthened. The surface
of the thong is studded with tubercles, from which
are discharged the seeds, accompanied with much
mucus, through the pores. The cup of this species
has been occasionally observed on exposed rocks,
swollen into a large hollow smooth black ball, exactly round, perhaps caused by the heat of the
sun rarefying and expanding the contained air, or
being perhaps the indication of a diseased state of
the plant.

A very remarkable form, and one of singular
beauty, is presented by the Peacock's tail (*Padina
pavonia*), a species not uncommon, attached to
rocks at the bottom of still, and generally shallow,
marine pools. The fronds rise in form of a rounded
fan, of a yellowish-olive tint, elegantly marked with
concentric zones or bands, of a dark brown. One
side, and sometimes both, is generally hoary, as if
dusted with powder, and the outer edge is delicately

THE PEACOCK'S TAIL   (*Padina pavonia*).

fringed with exceedingly minute filaments, which, in a living state, often reflect the prismatic colours of the rainbow.

Perhaps the most lovely of all the *Fuci* is the *De-lesseria sanguinea*, which is a common species. It consists of several oblong-oval or pointed leaves, of extreme delicacy, with the edges very much waved or plaited, furnished with a midrib and side-veins, which materially increase their leaf-like appearance; the colour is an exceedingly rich rose-pink. The midrib often throws out smaller leaves, which, if the main frond be destroyed, soon attains its usual size; an interesting provision against the accidents to which these apparently frail plants are neces-sarily exposed. The fructification of this genus is curious, as being of a twofold character : both forms are found in the winter, affixed to the mid-rib, which alone survives that season, the foliaceous part having all decayed away. The one mode is by means of nearly globular capsules attached to the rib by short foot-stalks, and inclosing many ir-regularly-shaped seeds; the other is by small mem-branaceous, leaf-like processes, likewise containing seeds. These two kinds of fructification occur on distinct individuals. This charming fucus, of which no adequate idea can be formed, by a verbal de-scription, retains much of its beauty when dried, and is very easily preserved. It is a pity that I am obliged to confess that its odour is very unpleasant, being rank and pungent.

Some of those species, whose fronds are very de-licately and numerously ramified, have been used to form mimic pictures. By skilful arrangement, very pretty landscapes are thus made, the forms and foliage of trees being beautifully imitated. The kinds most commonly appropriated for this purpose are *Plocamium coccineum* and *Gelidium cartilagi-neum*, which have a very beautiful effect if simply expanded on smooth white paper, or, on the pearly inner surface of large shells. The whole order *Flo-rideæ*, to which these belong, is remarkable for bril-liant hues, and often elegant forms.

Like their kindred, the plants of the earth and air, the sea-weeds have their parasites. As the *Tilland-sia* grows on the giants of the tropical forests, and as the mistletoe grows upon the apple-tree of our own orchards, so do some of these draw their nourish-ment, or at least derive their support, from the fronds or stalks of others. *Ptilota plumosa*, for example, a delicately-feathered species, of a pink or purplish hue, is found to be parasitical on the common tangle. It is justly considered one of the ornaments of our southern shores, but becomes still finer as we approach a more southern latitude. This must not be confounded with another elegant little plant bearing the same specific name, but belonging to a dif-

BRYOPSIS PLUMOSA.

ferent genus, *Bryopsis plumosa*. The tribe of which the latter is a member is remarkable for its delicacy: in the one now mentioned the main stem is very slender, set with horizontally-spreading branches, like a pine-tree, each of which is most elegantly feathered. Its colour is a bright grass-green, and the whole surface shines as if it were varnished. It is so delicate that in drying, the colouring matter contracts in the stem, leaving interrupted spaces destitute of colour, and perfectly transparent.

These are but a very few of the multitudinous sea-weeds which would come under the notice of an observant visitor to our own rocky shores; yet how manifold are the indications of infinite intelligence and goodness even in these things proverbial for their vileness! And while we gratefully acknow-ledge the Divine hand in such species as conduce to man's sustenance or comfort, may we not, from the lavish beauty and elegance of such as are of no direct benefit to us, legitimately draw the same consola-tory inference which the Saviour drew from the lovely lilies at His feet? If God so clothe these obscure caverns and submerged rocks, will He not *much more* care for those whom He has redeemed with the blood, and conformed to the image, of His Son? Nor is the relation which He sustains to these frail and perishing weeds limited to an exertion of creative power. All are marshalled in order, each is provided incessantly with the requisite supplies for its welfare, and each is assigned to that particular locality which suits its habit of growth, and where alone it flourishes. On this subject Mr. Neill observes, "On our open shores a certain order is observed in the habitat of the *Fuci*, each species occupying pretty regularly its own zone or station. *Chorda filum*, or Sea-laces, grows in water some fathoms deep: in places where the tide seldom en-tirely ebbs, but generally leaves from two to three feet of water, grow *Alaria esculenta* and *Laminaria* *bulbosa*, and the larger specimens of *L. digitata* and *saccharina*, with some small kinds, as *Rhodomenia palmata*, *Halidrys siliquosa*, and *Delesseria sanguinea*. In places uncovered only at the lowest ebbs, smaller plants of *L. digitata* and *saccharina* abound with *Himanthalia lorea*, or Sea-thongs. On the beaches uncovered by every tide, *F. serratus* occurs lowest down, along with *Chondrus crispus* and *mammillosus;* next comes *F. nodosus*, and higher up, *F. vesiculosus*. Beyond this, *F. canaliculatus* still grows, thriving very well if only wet at flood tide, though liable to become dry and shrivelled during a great part of the day. Lastly, *Lichina pygmœa* is satisfied if it be within reach of the spray."[*]

In examining these Algæ, and especially if we collect them for preservation, we shall find very frequently entangled among them, branches of a substance which adheres with so much tenacity as to cause no little trouble in cleansing the specimens. I refer to the common Coralline (*Corallina officinalis*). No organic substances have so much divided naturalists in opinion as to their real nature as the Corallines. Evidently placed on the very verge of the animal or vegetable kingdom, it required a minute acquaintance with their structure, derived from the closest observation, and all the research of modern science, to decide the long uncertain question, and to fix them where they now by common consent hold their place among the vegetable tribes. The one of which I speak, and the most

[*] Edin. Encyc. Art. "Fuci." Most of the species here alluded to I have described above.

common, being abundant on every rocky shore, ordinarily presents, though subject to much variation, the form of a spreading bushy tuft, from one to four inches high, growing from a broad stony base, of a shape more or less round. Each branch consists of many short joints, a little broader at the upper than at the lower end, which often send out other jointed branches from each upper shoulder, as well as from the centre. The joints are of a stony

CORALLINE (*Corallina officinalis*).

or rather shelly consistence, being chiefly a deposit of lime; when dead they are perfectly white, but in a living state they assume a purplish tint. Linnæus and many other eminent men were deceived by this shelly appearance into an opinion of their animal nature, maintaining that animals alone ever produced lime. But on removing the calcareous deposit, we perceive that it is merely a crust en-

F

veloping an axis of an evidently vegetable character.
On placing the Coralline in vinegar, or other weak
acid, the lime is dissolved, leaving the vegetable
part coloured as before, which, though continuous
through its length, is constricted at the parts which
corresponded to the joints of the crust, and looks
very much like one of the jointed *Fuci*. It is very
common to see the broad base without any jointed
branches, for the former attains some size before
the latter shoot, and may be seen in this state on
almost every object between the range of high and
low tide. It first appears as a thin, round, shelly
patch of a purplish hue, on the shell of a Mollusk,
or the frond of a *Fucus*, or the smooth rock,
and gradually enlarges by additions at the edge,
the progress of which is marked by concentric
zones, or rings of a paler tint, till it sometimes
attains several inches in diameter. It is tenacious
of vitality, and when the branches are all torn off
by the violence of the waves or other accidents,
the base still lives on, and becomes studded with
roundish knobs. This base, when growing on a
soft calcareous rock, will often increase much in
thickness, without showing any tendency to throw
out its jointed branches; or in situations where it
is long uncovered by the tide, and exposed to the
influence of the sun, it becomes "a softish white,
leprous crust." Its ordinary form, however, is by
far the most pleasing, particularly when growing, as
they delight to do, on the sides of the still, rocky
pools already described, their bushy tufts gracefully hanging over each other, like weeping wil-

lows in miniature. Beyond its beauty I know
not that this little creature has any obvious claim
to our consideration, except that, in common with
other sea-plants, it gives out oxygen, and thus
maintains the element in which it grows in a state
fit for the support of animal life. But this is a
service vastly important, and explains why the
"floor of the ocean" is covered, as it appears to
be, with such a profusion of vegetable life. And
here so wisely is the balance kept up between the
animals which absorb oxygen and the plants which
evolve it, that, perhaps, the world could not afford
to lose a single species of either without derangement of the existing order, which would be followed by manifest inconvenience. Of course our
little Coralline cannot do *much* to promote this
object; but that it does exert *some* beneficial influence, we have evidence in an experiment of
Dr. Johnston, whose researches on these neglected
tribes are so interesting. "Was there a need," he
observes, "of adding any additional proof of the
vegetability of the Corallines, an experiment in progress before me would seem to supply it. It is
now eight weeks ago since I placed in a small
glass jar, containing about six ounces of pure sea-
water, a tuft of the living *Corallina officinalis*, to
which were attached two or three minute *Confervæ*,
and the very young frond of a green *Ulva*, while
numerous *Rissoæ*, several little Mussels, and An-
nelides, and a Star-fish, were crawling amid the
branches. The jar was placed on a table, and was
seldom disturbed, though occasionally looked at;

and at the end of four weeks the water was found
to be still pure, the *Mollusca* and other animals all
alive and active, the *Confervæ* had grown perceptibly, and the Coralline itself had thrown out some
new shoots, and several additional articulations.
Eight weeks have now elapsed since the experiment was begun,—the water has remained un-
changed,—yet the Coralline is growing, and apparently has lost none of its vitality ; but the animals
have sensibly decreased in number, though many
of them continue to be active, and show no dislike to their situation. What can be more conclu-
sive? I need not say that if any animal, or even a
sponge, had been so confined, the water would long
before this time have been deprived of its oxygen,
would have become corrupt and ammoniacal, and
poisonous to the life of every living thing."*

Who is not familiar with Sponge,—with its-soft-
ness, its elasticity, its capacity of absorbing and retaining fluids, and other qualities which render it so
valuable in domestic economy? And yet how few
are aware that it is the skeleton of an animal! In
fact, Sponge is one of those dubious forms which
God has placed in the great system of Creation, on
the confines of the two great divisions of organic
beings, apparently having little in common with
either. Like the Corallines, the Sponges have afforded occasion for much controversy as to their
proper position; but they are now pretty unani-
mously assigned to the animal kingdom. The common Sponge of household purposes (*Spongia offici-*

* British Sponges, p. 215.

nalis) is a native of the Mediterranean, but is much more familiar to us than our native species, of which there are many. The appearance which it presents is that of an irregularly-shaped mass, more or less rounded, composed of a brown woolly substance, perforated by innumerable pores in all directions, and having in addition, wide canals communicating with each other, and terminating in round holes or mouths on the surface. But if we take a small portion of the substance, and place it under a common magnifying lens, we shall see that it is composed of shining, horny, nearly-transparent fibres, which, by uniting with each other at all angles and distances, form a loose and very irregular network. Now, when in a living state, every fibre was enclosed in a coating of thin, clear jelly, which formed the living animal, the horny fibres constituting, as I have intimated above, only the skeleton. Imbedded in the substance of many species, some British ones, for example, are found *spiculæ*, or needle-like crystals, of pure flint, varying much in shape in various kinds, while other species have similar crystals of lime. Where these occur in considerable numbers, the Sponge does not possess elasticity: it may be crushed, but it will not regain its original form. It is a singular fact, that Sponges of these three different kinds are sometimes found growing close to each other, and all alike nourished by the same simple fluid, pure sea-water; yet they elaborate therefrom products so different as horn, flint, and lime. The animal nature of Sponges is not easily to be detected: no indication of sensation has ever

been perceived in them when living, even though
violence in many modes has been offered to them;
though beaten, pinched with hot irons, cut or torn,
or subjected to the action of the strongest acids.
The substance may be destroyed, but there is no
contraction, nor the slightest evidence of feeling;
to all appearance they are as passive as the rock on
which they grow. One proof of their animality,
however, is open to any one: we are all familiar
with a peculiar smell produced when horn, wool,
feathers, &c., are burned; this smell arises from the
presence of *ammonia*, and is peculiar to animal matter; on burning a bit of Sponge this animal odour
is strongly perceptible. On viewing a living Sponge,
however, in water, with care and attention, it is
found to exhibit a constant and energetic action,
which sufficiently shows its vitality. Dr. Grant's
account of his discovery of this motion in a native
species is so interesting, that, though I have quoted
it in another treatise, I may be forgiven for repeating it here. "I put a small branch of the *Spongia
coalita*, with some sea-water, into a watch-glass,
under the microscope: and on reflecting the light
of a candle through the fluid, I soon perceived that
there was some intestine motion in the opaque particles floating through the water. On moving the
watch-glass, so as to bring one of the apertures on
the side of the Sponge fully into view, I beheld, for
the first time, the splendid spectacle of this living
fountain vomiting forth from a circular cavity an
impetuous torrent of liquid matter, and hurling
along in rapid succession, opaque masses, which it

strewed every where around. The beauty and novelty of such a scene in the animal kingdom long
arrested my attention; but after twenty-five minutes
of constant observation, I was obliged to withdraw
my eye from fatigue, without having seen the torrent for one instant change its direction, or diminish
in the slightest degree the rapidity of its course. I
continued to watch the same orifice, at short intervals, for five hours, sometimes observing it for a
quarter of an hour at a time; but still the stream
rolled on with a constant and equal velocity."

Sponges, in general, appear to have little choice
of situation, but to grow wherever the young offset
or gemmule happens to drop, whether on the rock,
on a shell, or on a sea-weed. If two of the same
species, growing side by side, come into contact,
their edges unite, and the two form one mass, so
perfectly one that the most practised eye could detect no indication of the line of union. On the contrary, if the neighbours be of different species, the
edges adhere by contact, but there is no union; and
both of the contiguous edges will grow up far beyond their natural level, like walls striving to over-
top each other, until the action of the waves prevents the continuance of a mode of growth so unnatural. Dr. Johnston speaks of two species of
Sponge which had become so intermingled in
growth, without being united, that, being of different colours, they presented the appearance of a
coloured map. The same writer has figured a much-
branched species (*Halichondria oculata*), growing on
the back of a small crab: the latter has a grotesque

appearance crawling under the perpetual shadow of
its own tree, the burden of whose weight, however,
was probably more than compensated by the protection it afforded against enemies.

A singular little creature, called the Hermit Crab
(*Pagurus*), the hinder part of whose body is unpro-
tected, except by a soft skin, is endowed with an
instinct which prompts it to seek some univalve
shells, into which it thrusts its abdomen, henceforth
using it as a house. Now there is a species of
Sponge found on our coast (*H. suberea*), of a corky
substance, which grows on the surface of similar
shells, overspreading and enveloping them; and it
so happens that in the great majority of instances,
the Sponge is found upon the individual shells inhabited by the Hermit. Gradually and insensibly
the Sponge grows over the shell, and at length creeps
round the edge of the lip, and begins to line the
inside: the constant motion of the crab, who is very
active, retards the growth for a while, but eventually
the Sponge prevails, and the Hermit, finding his pre-
mises becoming every day more and more contracted,
is at length compelled to seek another lodging. A
proceeding very similar to this, but which the Hermit Crab finds rather to his advantage than discomfort, takes place in the growth of a species of Coral
(*Alcyonium echinatum*). This coral also very frequently grows on a shell selected for a habitation
by the little crab; but as it grows, it does not line
the shell, but becomes moulded, as it were, to the
form of the enclosed animal, thus increasing the size
and commodiousness of the dwelling, and precluding

the necessity of quitting the tenement. Mr. Gray remarks on this:—"One can understand that the Crab may have the instinct to search for shells on which the coral has begun to grow; but this will scarcely explain why we never find the coral except on shells in which Hermit Crabs have taken up their residence."

One of the most pleasing forms that are presented by the Sponges, which are exceedingly various, is that of a cup with a dilated foot; it is about as large as a tea-cup, but is more funnel-shaped, whence its name (*H. infundibuliformis*). A similar species from the Indian seas, commonly called Neptune's Cup, though much larger, is inferior to our little goblet in neatness of appearance and sponginess of texture.

Our shores abound with examples of those astonishing forms of animal life, the Polypes, both simple and aggregated. The former under the names of Animal-flowers, and Sea-anemones, have attracted general admiration from their intrinsic beauty, and from their very close resemblance to composite flowers. When out of water, or reposing, they usually take a semi-globular shape, adhering by a broad base to the rocks, but some are somewhat lengthened and cylindrical. The centre of the upper surface is depressed, and there is evidently an aperture which has been closed. When seeking for prey this orifice opens, by its edges turning inside out, as it were, and dilates, until it is as wide as the base; while from within the lip, or outer rim, protrude a multitude of fleshy rays, called *tentacula*, arranged in three or four rows extending all round. In the

centre of the expanding disk is the real mouth, or
opening into the stomach. It is these tentacula,
which, spreading around exactly like the rays of
an aster or marigold, give to the Polype so striking
a likeness of a flower. These animals are exceedingly voracious; though when closed, you would
think them a mere lump of jelly-like flesh, utterly
helpless and incapable of any exertion; yet when
the tentacula are all expanded, no small crab, or
shrimp, or mussel, can even touch one of them with
impunity. From some cause, not thoroughly under-
stood, each tentacle has the power of adhering with
wonderful tenacity to any object on the slightest contact. I have often been surprised at the force required to draw away my finger when I have gently
touched one. No sooner, then, has some little
shelled Mollusk been thus caught, than instantly
other tentacles lay hold of it also, and it is inevitably
dragged by their contraction into the mouth. It
remains in the stomach a few hours, when the shell,
entirely cleared of all the meat, is vomited through
the mouth, there being but one orifice to the body.
The Polype is capable of great dilatation, which enables it to swallow an animal even much larger than
the ordinary dimensions of its own body. A very
curious instance of this I shall presently mention;
but first I must allude to that which forms the most
wonderful feature in its history, the power of repro-
ducing any parts that have been cut off. To so
great an extent does this power prevail, that even
if cut into many parts, each separate part will put
forth the parts wanting, and soon become a complete

animal. For example, if, with a sharp knife, a Polype be cut into two by a horizontal section, midway between the tentacles and the base, the upper portion will adhere to a rock, close the bottom of the stomach, and take its former shape; the under part will throw out rudimentary tentacles around the centre, which will soon be in a condition to take food, and the original form and functions will be displayed by this portion also. Nay, it has even been found that if, as often happens, the animal, being violently removed from its support, leave behind any fragments of its base still adhering, each of these torn portions will, in a short time, acquire all the parts of the perfect animal. These powers strongly remind one of vegetable life; for it is as if one were making cuttings, and consequently new plants, of a fuchsia or verbena. The ordinary mode in which the Polypes continue their race is very plant-like; the young grow from any part of the surface like little buds, and when they have attained the form of the parent, drop off; often, however, they are vomited through the mouth. Any of my young readers who live near the coast may easily verify these observations; but I would not recommend the artificial mode of increasing the animals, because, though it may well be doubted whether they are susceptible of pain, such experiments have an appearance of cruelty at least, which it is well to avoid. In some situations you will find in abundance *Actinia gemmacea*, the most lovely of our native animal flowers, which I will describe. When closed, it is of a rounded or sometimes oval

shape, somewhat flattened, about an inch and a
half in diameter, very variable in colour: sometimes being of a brilliant scarlet with pale warts,
like rows of ornamental beads; at other times it
is of a sulphur yellow, or pale green, with stripes
of orange colour; and I have seen specimens of
a lively rose-pink, studded with green dots. When
expanded, it displays three or four circles of tentacles, which are rather short and thick, and varie-
gated with white and red in alternate rings.
Sometimes, by imbibing a large quantity of water,
it becomes distended to twice its usual dimensions,
and is then nearly transparent. There is an instinct displayed by this species, which one would
not expect to find in a creature of so low an organ-
ization, and which is worthy of our admiration,
as showing how mindful the gracious Creator and
Preserver is of His creatures' well-being. Such
individuals as have taken up their residence upon
the half-submerged rocks, where the daily recess
of the tide exposes them to observation, are covered
with rough warts, and blotched with dusky brown
and dull orange, and are coated with fragments of
shells, sea-weed, and gravel, which adhere to the
skin by a glutinous secretion, so strongly as not
to be washed off; and being thus veiled, the animals defy detection. On the other hand, those
specimens which live in deep water, as if aware
that the necessity for concealment no longer exists, have nothing of the kind, their skins are
smooth and naked, and adorned with the vivid
tints which make this species so beautiful. The

Actinia are easily procured, and kept alive a long time in sea-water without difficulty; in a glass vessel their beauty is displayed to advantage, needing only the precaution of supplying them with pure sea-water every two or three days at most, or they will throw off their skin in ragged pieces, become discoloured, and die. They are capable of very long fasts, although, as I observed before, voracious enough when food is to be obtained. Dr. Johnston tells us of a specimen of the *A. gemmacea* once brought to him, "that might have been originally two inches in diameter, and that had somehow contrived to swallow a valve of *Pecten maximus* (the great Scallop) of the size of an ordinary saucer. The shell, fixed within the stomach, was so placed as to divide it completely into two halves, so that the body, stretched tensely over, had become thin and flattened like a pancake. All communication between the inferior portion of the stomach and the mouth was of course prevented; yet, instead of emaciating, and dying of atrophy, the animal had availed itself of what undoubtedly had been a very untoward accident, to increase its enjoyments and its chances of double fare. A new mouth, furnished with two rows of numerous tentacula, was opened upon what had been the base, and led to the under stomach : the individual had indeed become a sort of Siamese twin, but with greater intimacy and extent in its unions!"*

Each of these animal flowers, except in the case of such accidental monstrosities as the one just men-

* Brit. Zooph. p. 224.

tioned, is a distinct and independent animal; but
there are some which, while they possess a general
similarity in structure to these, exist only in aggregated communities; many individual Polypes being
clustered upon a somewhat solid body called a Polypidom, which is, when alive, clothed with a fleshy
coat, believed to be capable of communicating and
receiving sensations to or from all the Polypes.
The teat-shaped bodies, familiarly called by the
fishermen Cow's-paps, when simple, and Dead-man's
toes, when branched, is a common example; the
Alcyonium digitatum of zoologists. It consists of
a leathery substance, capable of contraction, studded
with orifices, whence project little stars with eight
rays, which are the expanded tentacles of the small
Polypes that inhabit the hollows. Those beautiful
productions, the Corals, some of which I may have
occasion to notice hereafter, are also formed on the
same model. They have generally a more solid
stem, partaking of the nature of stone, and branch
out in imitation of shrubs. The stony or horny
centre is, however, clothed with gelatinous flesh, in
which, as in the former instance, hollows occur at
intervals, occupied by minute star-shaped Polypes.
The warty white coral (*Gorgonia verrucosa*), not
uncommon with us, is of this structure, .having a
stony skeleton; but in the beautiful Sea-fan (*G.
flabellum*), the skeleton shows more the texture of
bone, or perhaps of horn; it is black, but is clothed
with flesh of a yellow colour, or sometimes purple.
From the ramifications being very numerous, and
uniting with each other at short intervals, like the

meshes of a net, this species is a very beautiful one. Its polypes, as in the other instances, have eight tentacles. This is exceedingly rare, though it has occurred on the British shores.

But more singular than either of these is the form of Polypidom, often brought up by fishermen attached to their baits, and by them called Cocks'-comb, or rather more appropriately, Sea-pen (*Pennatula phosphorea*). It very closely resembles a

SEA-FAN (*Gorgonia flabellum*), and SEA-PEN (*Pennatula phosphorea*).

broad feather from two or three inches in length, and of a purplish colour. The lower part is cylindrical, or nearly so, and represents the quill, and the tip of this is tinged with orange. Above this the stem is fringed on each side with very regular, flat, dentated processes, diminishing gradually

to the tip, representing the vane. Along the upper
edge of each of these *pinnæ* are placed the cells,
inhabited by minute, white, eight-rayed Polypes.
The stem contains a long, needle-shaped bone, very
slender at each extremity, which is bent backwards
so as to form a hook. Some authors have affirmed
that the Sea-pen swims freely in the sea by the
waving motion of its *pinnæ;* but modern observations tend to throw discredit on this statement,
which in itself seems improbable: the fishermen
affirm that it abides with its stem inserted in the
mud at the bottom; and those which have been
kept for observation have remained at the bottom
of the vessel, without any apparent power of even
turning over on the other side. This species, as
its scientific name imports, is one of the many animals that inhabit the sea, which are endowed with
the faculty of producing light: in this instance, it
appears from experiments that the power is exerted
as a means of defence, as only when injured or irritated does the animal give out its light, which is of
a faint-bluish cast. Its sudden illumination at the
bottom of the sea may have the effect of terrifying
some of its enemies, and of thus protecting it from
the dangers to which its otherwise helpless frame
would be exposed.

THE SHORES OF BRITAIN.

CONTINUED.

THERE is one aspect in which, if we view the sea,
it speaks eloquently the beneficence of God to man;
namely, as the source from whence he draws an inexhaustible supply of wholesome and nourishing food.
And there is no nation more favoured in this respect
than Great Britain: the seas which surround us are
stocked with a vast variety of fishes, the great majority of which are eatable. From the form of our
coasts, there is always at some part access to the
sea, the wind which locks up the ports of one coast
leaving others free; the numerous bays, harbours,
and inlets offer a refuge to which to run in unfavourable weather, as well as a market for the dis-
posal of the produce taken; while the bold and
hardy character of our population qualifies them to
take advantage of a proffered source of profit, though
not unattended with risk. Accordingly, we find
that the fisheries afford to this country a revenue
of great value; and an immense quantity of cheap
animal food is produced by them, the importance of
which can hardly be overrated. The prosperity of
Holland is notoriously founded upon the zeal, industry, and success with which her sons have prosecuted the herring-fishery; a fact which is announced

in the well-known Dutch saying, "The city of Amsterdam is built upon herring-bones:" and though, from the superiority of our internal resources, we are not compelled to give so undivided an attention to the scaly tenants of the deep as they have been, we may still assert, that on a similar base stand many of our important seaport towns. Let us then examine these finny tribes, which come so strongly recommended to our notice, and see if we cannot discover in their formation and economy evidences of that all-pervading wisdom and goodness of which we have had occasion before to speak.

An intelligent observer can scarcely fail to be struck with the perfect adaptation of fishes for swift motion through a dense fluid. The form most suited for rapid progression is that of a spindle, swelling in the middle and tapering to the extremities: and this is the general form of fishes. The variations from this normal shape are comparatively rare, and consist chiefly in the lengthening of the body, as in the Eels, or in widening its diameter perpendicularly, as in the Flat-fishes, or horizontally, as in the Skates. But in these cases, and similar ones, the exceptions are made to suit variations in habits, for the Skates and Flat-fishes are intended not for rapid swimming, but for lying flat upon the bottom; while the worm-like form of the Eels enables them to insinuate themselves with facility through the mud and ooze, or even to leave the water and crawl upon the shore. Still, however, in both the usual form is to be traced, the central part of the body being the widest and the extremities being pointed. The facility of motion possessed by fishes is partly dependent on their simplicity of figure, the absence of those prominent limbs which project from the bodies of most other vertebrate animals; the head, without any visible neck, merging into the rounded body, which terminates in the tail in an almost unbroken outline, for the fins are usually so slight and membranous in their texture as scarcely to diminish this unity of form. The smooth and glittering armour, in which these animals are for the most part invested, tends to the same end. Feathers or fur would greatly impede progress through water; and as the tribe of fishes are what is commonly called cold-blooded, or of nearly the same temperature as the fluid that surrounds them, those non-conductors of heat would be of no service, the animal heat necessary for existence not being liable to be abstracted. In place of those clothing substances, the fish's body is encased in a coat of mail formed of many pieces of similar shape, of a transparent horny substance, which are imbedded in the skin on the side next the head, and overlap the succeeding ones at the posterior edge, like the tiles of a house. It is obvious how beautifully and effectually this formation precludes any impediment in swimming, arising from the free edges of the scales. These are so closely pressed on each other, that the water cannot penetrate, and are covered, moreover, in many fishes with a glutinous slime, which water does not dissolve. The scales of fishes afford objects of very beautiful structure when viewed with a microscope. They are various in their form; those from different parts of the body not being quite alike even in the same fish. They are not perfectly flat, but take the form of a very flattened cone, of which the apex is usually a little behind the middle. Between this point and the edge there is a great number of concentric flutings, too fine, as well as too near each other, to be readily counted; and it is presumed that each of these lines indicates a stage in the growth of the scale; that the scale is increased, perhaps annually, or perhaps oftener, by a deposit of horny matter on the surface next the skin, each of which deposits exceeds in diameter that which preceded it, by a very minute amount on every

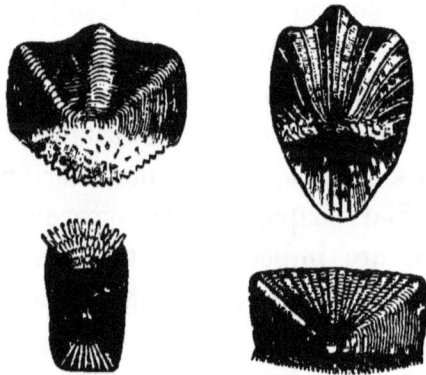

SCALES OF FISHES.

side. The concentric lines are often traversed by other lines, diverging with great regularity from the apex. The edges are sometimes cut into points, scallops, or waves, of exquisite symmetry; the surface is often variously sculptured; and the whole presents a specimen of the most elaborate workman-

ship, worthy of the Divine hand that formed it.
The scales of some fishes are so minute as to be
with difficulty distinguishable; such as those of the
Eel: to procure these for microscopical examination,
"take a piece of the skin of the Eel that grows on
the side, and while it is moist spread it on a piece
of glass, that it may dry very smooth; when thus
dried, the surface will appear all over dimpled or
pitted by the scales, which lie under a sort of cuticle,
or thin skin: this skin may be raised with the sharp
point of a penknife, together with the scales, which
will then easily slip out, and thus you may procure
as many as you please."*

The limbs of fishes differ greatly in appearance
from those of terrestrial animals; consisting, as to
the portion external to the body, of slender spines,
sometimes cartilaginous and jointed, at others bony
and simple, united by means of a thin membrane
stretched from one to the other. Generally there
are two pairs on the under part of the body, which
are called the pectoral and the ventral fins, and represent respectively the fore and hind legs of qua-
drupeds, or the wings and feet of a bird. Besides
these, there are one or more perpendicular fins along
the back, called the dorsal, and one below the body,
near the tail, called the anal; but the main instrument of motion is the broad, perpendicular fin, which
terminates the body, often called the tail, but, more
correctly, the caudal or tail fin. To rightly understand the motions of a fish, we must bear in mind
that it is immersed in a fluid which is of little less

* Martin's Micrographia Nova, p. 29.

6

specific gravity than its own body; but in order to
regulate its own weight, it is provided with an internal bladder, filled with air, and furnished with
muscles for its compression or expansion: by the
former process rendering its body heavier, and by
the latter lighter than the water. It is true there are
many fishes which are destitute of the air-bladder;
but these are, for the most part, ground fishes, which
reside habitually upon the bottom, rarely swimming
to any distance. The tail, as was observed, is the
grand organ of progression; and most of the muscles
of the body are so inserted upon the joints of the
spine as to give the greatest possible energy to the
motions of this organ. Its expansion is vertical, and
its motion is only horizontal, from right to left: so
that, striking the water on either side with great force,
the fish shoots rapidly forward in the direction of
the line of the body, but cannot, by its means, ascend
or descend. The direction of a fish's motion is governed by the pectoral and ventral fins, which aid,
likewise, in balancing the body, and obviate the
tendency to turn belly uppermost, a position which
a dead fish assumes, from the weight of the muscular
back being superior to that of the hollow and air-
filled belly. There is considerable diversity in the
depth of water which different species of fishes habitually inhabit; and this depends, in a great measure, on
the position of the ventral fins. Such as mainly reside
at or near the surface have them so placed that the
centre of the body shall fall nearly midway between
them and the pectorals. Those whose habits lead
them to range to great distances without any material

change in their depth of water, have the ventral fins placed far back on the belly, as the Herring and the Salmon; while those which feed at the bottom in deep water, but yet have considerable power of swimming, such as the Cod, require the ventrals to be situated near the head, sometimes even in advance of the pectorals, in order to act with rapidity and effect upon the fore part of the body, which is usually heavy in such fishes. The Flat-fishes and Skates, in which the ventrals are little developed, and the Eels, in which they are wanting, rarely quit the ground, but grovel on the mud in shallow water. Many fishes have certain spines of the fins developed into stiff and formidable weapons, and others have equally effective armour placed upon the gill-covers, the sides of the body or the tail. With these, which are usually jointed, and which the fish has the power of erecting stiffly, and of directing with considerable precision, it sometimes inflicts severe wounds on the incautious fisherman, as well as on its opponent, in the battles with its own kind, which often occur. The little Stickleback (*Gasterosteus*), which abounds all round the coast, as well as in our fresh waters, is armed with sharp spines on the back and sides, which it wields like a perfect tyrant. "When a few are first turned into a tub of water, they swim about in a shoal, apparently exploring their new habitation. Suddenly one will take possession of a particular corner of the tub, or, as it will sometimes happen, of the bottom, and will instantly commence an attack upon his companions; and if any one of them ventures to oppose his sway, a regular and most furious

battle ensues; the two combatants swim round and
round each other with the greatest rapidity, biting,
and endeavouring to pierce each other with their
spines, which on these occasions are projected. I
have witnessed a battle of this sort which lasted
several minutes before either would give way; and
when one does submit, imagination can hardly conceive the vindictive fury of the conqueror; who, in
the most persevering and unrelenting way, chases his
rival from one part of the tub to another, until fairly
exhausted with fatigue. They also use their spines
with such fatal effect, that, incredible as it may appear, I have seen one during a battle absolutely rip
his opponent quite open, so that he sank to the bottom, and died. I have occasionally known three or
four parts of the tub taken possession of by as many
other little tyrants, who guard their territories with
the strictest vigilance, and the slightest invasion invariably brings on a battle."* The Sting-rays (*Try-
gon*), which are furnished with a hard and sharp spine
with toothed edges, near the base of the tail, are accustomed to twist their long and flexible tail around
their enemy, while they inflict severe wounds with
the barbed spine. The Common Skates (*Raia*), on
the other hand, which have the tail studded with
rows of curved horny thorns, when irritated, are said
to bend the body nearly into a circle, and to dash
about the armed tail with violence in all directions.

The goodness of God is manifest in the gregarious
habits of most of those fishes which constitute an important article of human food, in the innumerable

* Mag. Nat. Hist. iii. 329.

individuals of which the shoals are composed, and in the fecundity by which the populousness of these shoals are maintained. Nine millions of eggs have been ascertained to exist in the roe of a single Cod, and the hosts of this, and other species, which during the fishing-season crowd our shores, are utterly beyond human calculation. These swarms were formerly believed to perform vast annual migrations in military order from the Polar regions in spring, and back again to their homes " beneath the ice" in the autumn. The groundlessness, and even absurdity of this notion has been shown, and it is now generally known, that the fishes are at no part of the year more than a few miles distant from the coast, but that on the approach of warm weather an unerring instinct teaches them, as by common impulse, to seek the shallows near the shore, in order to deposit their spawn within the vivifying influence of the summer sun. This grand business of life being accomplished, they again retire, not to the Arctic ice, but to the deep water of the offing, where they may again rove in freedom and conscious security. And this is an admirable ordination of Divine Providence, that these tribes are thus periodically brought within the reach of man precisely at the season when they are in the highest condition, and therefore most wholesome, as well as most agreeable. For they come from the deep water fat, and in full health and vigour; but after having spawned they return sickly and poor, to recruit their exhausted strength.

The Herring family (*Clupeadæ*), including the

common Herring, the Pilchard, the Sprat, the Shad,
&c., are the most important objects of our fisheries,
and particularly the first-named two species.

The fishery for the Pilchard is carried on almost
exclusively in the counties of Cornwall and Devon;
the Herring is more generally diffused, but the
greatest numbers taken are on the shores of Scotland and the adjacent islands. Some idea of the
commercial importance of these two animals may be
formed from the facts, that between three and four
hundred thousand barrels of Herrings are sometimes
cured in a single year in Great Britain alone, besides
all that are sold while fresh; and that ten thousand
hogsheads of Pilchards have been taken on shore
in one port in a single day, "thus providing," says
Mr. Yarrell, "the enormous multitude of twenty-five
millions of living creatures drawn at once from the
ocean for human sustenance." The shoals of Herrings
are occasionally known to approach the shore with so
headlong an impetuosity as to be unable to regain
deep water, and are stranded upon the beach in immense numbers. Mr. Mudie has described such an
incident. "The rocky promontory at the east end
of the county of Fife, off which there lies an extensive reef or rock, sometimes has that effect, and there
have been seas [seasons?] in which, when the difficulties of the place were augmented by a strong wind at
south-east, that carried breakers upon the reef, and a
heavy surf along the shore, the beach for many miles
has been covered with a bank of Herrings several
feet in depth, which, if taken and salted when first
left by the tide, would have been worth many

thousands of pounds, but which, as there was not a
sufficient supply of salt in the neighbourhood, were
allowed to remain putrefying on the beach until the
farmers found leisure to cart them away as manure.
One of these strandings took place in and around the
harbour of the small town of Crail only a few years
ago. The water appeared at first so full of Herrings
that half a dozen could be taken by one dip of a
basket. Numbers of people thronged to the water's
edge, and fished with great success; and the public
crier was sent through the town to proclaim that
"caller herrin," that is, Herrings fresh out of the sea,
might be had at the rate of forty a penny. As the
water rose the fish accumulated, till numbers were
stunned, and the rising tide was bordered with fish,
with which baskets could be filled in an instant. The
crier was, upon this, instructed to alter his note, and
the people were invited to repair to the shore, and
get Herrings at one shilling a cart-load. But every
successive wave of the flood added to the mass of
fish, and brought it nearer to the land, which caused
a fresh invitation to whoever might be inclined to
come and take what Herrings they chose gratis. The
fish still continued to accumulate till the height of
the flood, and when the water began to ebb, they
remained on the beach. It was rather early in the
season, so that warm weather might be expected;
and the effluvia of many putrid fish might occasion
disease; therefore the corporation offered a reward
of one shilling to every one who would remove a full
cart-load of Herrings from that part of the shore
which was under their jurisdiction. The fish being

immediately from the deep water, were in the highest
condition, and barely dead. All the salt from the
town and neighbourhood was instantly put in requi-
sition, but it did not suffice for the thousandth part
of the mass, a great proportion of which, notwithstanding some not very successful attempts to carry
off a few sloop-loads in bulk, was lost."*

The Herring appears on our shores in the middle
of summer, but seems to approach the coast of Scotland earlier; for in Sutherland the fishery commences
in June, and in Cromarty even so early as May,
while the Yarmouth season rarely begins till September. They are taken chiefly by means of drift-nets,
and by far the majority are cured: in the first part
of the season, however, they are often so rich as to be
unfit for salting, and these are sold for consumption
while fresh. About the month of November the
shoals spawn, and are then unfit for eating, and the
fishery ceases. As is universally known, there are
two modes of curing this fish, producing what are
called white and red herrings. The former requiring
only to be placed in barrels with salt, the process can
be performed in the fishing-craft; consequently the
vessels for this fishery are larger, being qualified to
keep the sea. Red herrings, however, require a
much more elaborate process, which cannot be performed on board, and the procuring of them is essentially a shore fishery. The Yarmouth men confine
themselves to this branch. They sprinkle the fish
with salt, and lay them in a heap on a stone or brick
floor, where they remain about six days; they are

* Brit. Naturalist.

YARMOUTH JETTY, IN THE HERRING FISHERY.

then washed, and spitted one by one on long wooden rods, which pass through the gills; great care is required that they may not touch each other as they hang; the rods are then suspended on ledges, tier above tier, from the top of the house to within eight feet of the ground; a fire is then kindled and fed with green wood, chiefly oak or beach, and maintained with occasional intermissions, for about three weeks, or, if the fish are intended for exportation, a month; the fire is then extinguished, and the house allowed to cool, and in a few days the herrings are barrelled.

2 H

Next in importance to the members of the above valuable family is the Mackerel, the most elegantly beautiful of the finny tribes that throng our shores. It is in season earlier than the Herring, usually appearing in spring, and the fishery is prosecuted in May and June, as in the latter month it spawns. It occurs in most abundance in the southern part of the kingdom, the coasts of Kent and Sussex being the chief stations of the fishery. The Mackerel is taken principally by nets, which are so set as to arrest the fish while roving about during the night; many, however, are taken by means of the hook, the favourite bait being a strip of flesh cut from the tail of a fresh Mackerel, or, in default thereof, a bit of red cloth: the fish bite most readily when the boat is sailing rapidly before the wind. The value of this fish depends, in a more than common degree, on its freshness; and hence it is important that no time be lost in conveying it to market. Fast-sailing boats are therefore kept in readiness to convey the cargoes to London as soon as caught, which usually find it advantageous to secure the aid of steam in ascending the river, as the loss of a single tide may diminish the value of the cargo one half, or even render it utterly unsaleable. During the season, not less than one hundred thousand are thus brought to Billingsgate per week.

The preceding species, coming in swarming shoals into the shallow waters, are usually taken by nets; but the Cod, another very valuable fish having different habits, is taken singly, by hook and line. It does not appear that the Cod is gregarious from

MACKEREL-BOAT OFF HASTINGS.

choice; or in any other sense than that of many
individuals independently actuated by a similar mo-
tive, flocking to any place where food is plentiful.
The Cod rarely comes into the shallows; but haunts
the deep water, feeding at the rocky bottom, on
marine worms, crustacea, and shelled mollusca. It
is a voracious fish. Mr. Crouch records having taken
thirty-five crabs, none of them less than a half-crown
piece, from the stomach of a single Cod: his greediness is often his own destruction and the fisherman's advantage, for it induces him readily to seize
the bait. It is most abundant on the north and
west coasts of Scotland, but is taken in considerable plenty all round the coasts of our island. In

some of the Hebrides there are large pools for the
preservation of sea-fishes, hollowed out of the solid
rock, and communicating with the sea by narrow
clefts at high tide. Great numbers of Cod-fishes
are kept in these vivaria, and are fed with various
garbage, or the bodies of other fishes. The stock
is replenished by casting in such individuals as are
but slightly injured by the hook in fishing, while
small ones, or such as are lacerated, are thrown into
the same receptacle, as food for their more fortunate
brethren. There are two modes of capturing the
Cod with the hook : the one is with what are called
in Cornwall bulters, which are long lines, to which
are attached, at regular distances, other lines six feet
in length, each bearing a hook; the intervals are
twice the length of the small lines, to prevent their
intertwining; these are shot across the course of the
tide. The other mode is by hand-lines, of which
each fisherman holds two, one in each hand, and
each line bears two hooks at its extremity, which
are kept apart by a stout wire going from one to the
other. A heavy leaden weight is attached near
the hooks, and thus the fisherman feels when his
bait is off the ground. He continually jerks them
up and down, and is thus aware of a fish the moment
it is secured. Although this seems a somewhat
tedious process of fishing compared with the immense draughts of the net, it is found in skilful
hands to be productive : eight men on the Dogger-
bank have taken eighty score of Cod in a day. It
is a heavy fish : Pennant records one which weighed
78lbs., but this was a giant; it was sold at Scar-

borough for one shilling! The fish are brought to
the mouth of the Thames in stout cutters, furnished
with wells, in which they remain alive; hence they
are sent up in portions to Billingsgate by the night
tide. The cutters lie at Gravesend: for if they
were to advance any higher up the river, the ad-
mixture of fresh water would kill the fish in the
wells. The liver of the Cod is not the least valuable part of its body, because it melts almost
entirely away into a clear oil, much used in manu-
factures.

There is a family of fishes familiar to us, which
are worthy of a moment's notice, not only on account of their importance as objects of commercial
speculation, but for their singular and unparalleled
deviation from the ordinary structure. These are
the Flat-fishes (*Pleuronectidæ*), comprising the Turbot, Plaice, Sole, and some others. Their form is
very deep, but at the same time very thin, and they
are not constituted to swim as other fishes do, with
the back uppermost, but lying upon one side. They
reside wholly upon the bottom, shuffling along by
waving their flattened bodies, fringed with the dorsal
and anal fins; and as they are somewhat sluggish in
their movements, they need concealment from enemies. This is afforded to them by the side which
is uppermost being of a dusky-brown hue, undistinguishable from the mud on which they rest; and
so conscious are they where their safety lies, that
when alarmed, they do not seek to escape by flight,
like other fishes, but sink down close to the bottom,
and lie perfectly motionless. Even the practised

eye of the turbot-fisher, with his powers sharpened
by interest, fails to detect a fish when thus concealed; and he is obliged to have recourse to another
sense, tracing lines upon the mud with an iron-
pointed pole, that the touch may discover the latent
fish. In the structure of the head, again, there is
a peculiar and very remarkable provision for the

TURBOT-BOAT OFF SCARBOROUGH.

wants of the creature. If the eyes were placed as
in all other animals, one on each side of the head,
it is plain that the Flat-fishes, habitually grovelling
in the manner described, would be deprived of the
sight of one eye, which being always buried in the
mud, would be quite useless. To meet this difficulty, the spine is distorted, taking, near the head,
a sudden twist to one side; and thus the two eyes

are placed on the side which is kept uppermost, where both are available. The inferior side of a Flat-fish is always white. The Turbot is the most highly esteemed of this family, and perhaps of all our fishes, the flesh being of very delicate flavour. The Sole is also a valuable fish. Both of these species are taken chiefly by trawl-nets, but the former is also caught with the hook.

The Crustaceous and Testaceous classes afford employment to a considerable number of our population, and demand our brief attention. Of the former, the chief species selected for food in this country are, the Crab, the Lobster, the Prawn, and the Shrimp. Both our salt and fresh waters, however, contain multitudes of other species, some of which are exceedingly curious in structure and form. The *Crustacea*, like insects, have no internal skeleton; but instead of it, are encased in a jointed framework, resembling the plate armour of our forefathers, of a texture between shell and bone. The muscles which move the body are attached to the interior of this crust, as our muscles are attached to the bones. The body consists mainly of two parts; the fore-division contains the head and chest, covered with a large single plate, and the hinder, the belly covered with several smaller plates, joined by a tough skin, and lapping over each other. As this shelly covering is possessed by the animal from its very birth, it is natural to inquire how it can possibly increase in size, seeing it is enclosed in an unyielding prison. In the Tortoises, which are somewhat similarly encased, the difficulty is met

by a periodical addition to the interior surface of
every plate a little wider in diameter than the one
before, thus enlarging the capacity of the aggregated plates, together with the enlargement of each
plate; and this, as I have already observed, is the
mode by which the scales of a fish grow. But from
the shape and size of the plates on a Crab or a
Lobster, and especially of the great one that en-
velops the chest, this mode of growth would not
answer the purpose. Another contrivance is resorted to, of a character perfectly unique; one of
those contrivances that meet us at every turn in
the study of Nature, and that make it so interesting and instructive, as manifesting the infinite resources of the Mighty God. When the Crustacean
finds that from its increasing size it is bound and
pressed by its shelly covering, it retires to some
hole or cranny for protection, becomes sickly, and
refuses to eat. After pining awhile, the softer
parts separate from the inside of the crust, even
the muscles becoming detached from the skeleton,
and take up a much smaller bulk than before: a
thick skin forms over this soft body, replacing the
crust, and then the great shield of the chest is
thrown off unbroken, and the other plates of the
body follow. This seems plain: but it is not so
easy to understand how the process is completed.
Every one who has looked at a Crab's claw, knows
that in a healthy animal it is filled with flesh, that
the inside is capacious, but that the joints are very
small: now, how is the animal to get its flesh freed
from this capacious boot? One would readily say,

by splitting it into two portions; but on examining the cast-off claws, which are frequently met with, no split or separation can be discovered. The question is not yet satisfactorily solved; but I believe that through the wasting away of the limbs from sickness and fasting, they become so diminished in size as to be drawn even through the narrow orifices of the joints. Every part of the old shell being thus thrown off, antennæ, eyes, jaws, and all, the animal fills its body with water, dilating all the parts to a size much exceeding that of the old shell, which the new skin, yet soft and flexible, readily permits. It is necessary that this inflation of the body should take place when newly freed, because the skin immediately begins to grow rigid, by lime being deposited in its substance secreted within the body, and rapidly takes the texture and consistence of the shell just rejected. The appetite now returns, and abundance of food soon restores the enlarged animal to its wonted vigour.

The Crabs, of which there are many species, have the shield of the chest very large and flat, and usually wider than long: the plates of the belly are small, and folded under the body out of sight. The great pincers or claws have considerable muscular power, and are covered, especially at the extremities, with a shell of almost stony hardness. The Crab wields these formidable weapons with much dexterity, and if he obtains a grasp, holds his opponent with persevering tenacity, so that he is not to be despised in single combat. Mr. Mudie tells an amusing anecdote illustrative of this habit. "We remember,"

says he, "an instance in which, but for timely assistance, the corporation of a royal borough would have
been deprived of its head, through the retentive
clutching of a Crab. The borough alluded to is
situated on a rocky part of the coast, where shell-fish
are so very abundant that they are hardly regarded
for any other purpose than as bait for the white
fishery. The official personage was a man of leisure;
and one favourite way of filling up that leisure was
the capture of Crabs, which, after much care, he had
learned to do by catching them in the holes of the
rocks, so adroitly, as to avoid their formidable pincers. One day he had stretched himself on the top
of a rock, and thrusting his arm into a crevice below,
got hold of a very large Crab; so large, indeed, that
he was unable to get it out in the position in which
it had been taken. Shifting his position in order to
accommodate the posture of his prey to the size of
the aperture, he slipped his hold of the Crab, which
immediately made reprisals by catching him by the
thumb, and squeezing with so much violence, that
he roared aloud. But though there be a vulgar opinion, of course an unfounded one, that Lobsters are
apt to cast their claws, through fear, at the sound of
thunder or of great guns, the thundering and shouting of the corporation man had no such effect upon
the Crab. He would gladly have left it to enjoy its
hole; but it would not quit him, but held him as
firmly as if he had been in a vice; and though he
rattled it against the rocks with all the power that
he could exert, which, pinched as he was by the
thumb was not great; yet he was unable to get out

of its clutches. But, 'tide waits for no man,' even though his thumb should be in a Crab's claw; and so the flood returned, until the greater part of the arm was in water, and the ripple even beginning to mount to the top of the rock, which, as the tides were high at that particular time, was speedily to be at least a fathom under water; and destruction seemed inevitable. A townsman who had been following the same fishery with an iron hook at the end of a stick, fortunately came in sight; and by introducing that, and detaching the other pincer of the Crab, which is one of the common means of making it let go its hold, he restored the official personage to land and life."*

The fisherman, however, prefers another mode of taking Crabs, than by seeking them in their rocky retreats. He uses pots made of wicker-work, with an opening in the top, made by the ends of the rods, bent inwards, and converging towards a point; their elasticity allowing a Crab to enter readily enough, but causing them to spring back to their first position when he is in, presenting only their converged points when he wishes to escape; the entrance being in the top of the pot, moreover, he cannot well get at it when once inside. Some decaying animal matter is put in by way of bait, which is an unfailing temptation to the Crab's palate, and the pot is sunk in deep water by means of a heavy stone. A line attached to a float on the surface of the water, marks the situation of each pot, and prevents mistakes as to property.

* Brit. Naturalist, i. 279.

CRAB-POTS.

The Lobster is caught in the same manner as the Crab, and both are in great demand for the delicacy of their flesh. A very large proportion of those eaten in England are brought from Norway. At first there does not seem much in common in the form of these two animals, except that both are furnished with pincers; but on examination, we shall find that both are constructed on the same model. The shield of the chest, which was broad and flat in the Crab, is long and arched in the Lobster; and the belly, which was thin, small, and folded out of sight, under the body, is in the latter much larger, and though bent, may be extended, and is terminated by fringed horny plates like a fin; the antennæ, or horn-like processes of the head, are very long. Thus we perceive, and there are many other examples which might be adduced, that it has pleased God to vary the forms of created beings, not by making each on a separate and independent plan, but by creating certain forms, which are viewed as types or models, and varying the parts, common to many species, in detail. The one mode would have been as easy as the other; there can be no gradations of facility in creation to Omnipotence; but doubtless He had wise ends in view in thus proceeding, though we may fail, from ignorance, in discerning them. Probably one reason may have been the formation of one harmonious whole out of the multitude of living creatures, which could not have been formed had every one been essentially different from all others. But, as it is, we see that deviations in structure and form are gradual, that one species varies but little from a certain type, another varies a little more, and so on; thus connecting each with each in a most beautiful order, something like the manner in which the links of a chain hang from each other, or perhaps still more, like an immense number of circles, so arranged as to touch other circles in many parts of their circumference. Goldsmith flippantly asserts, that the Shrimp and the Prawn " seem to be the first attempts which Nature made when she meditated the formation of the Lobster." Such expressions as these, however, are no less unphilosophical than they are derogatory to God's honour; these animals being in an equal degree perfect in their kind, equally formed by consummate wisdom, incapable of improvement, each filling its own peculiar place in its own circle, which the others could not fill.

THE SHRIMPER.

The Shrimp and Prawn, like the Lobster, have the extremity of the body furnished with broad overlapping plates, strongly fringed, which, expanding in the shape of a fan, constitute a powerful fin. The body, a little behind the middle, has a remarkable bend downwards, though it may be brought nearly straight. Their motion when swimming is very swift, and in a backward direction, and is performed by striking the water forcibly with the tail-fin, the body being in a bent position. The Lobster is said to project itself thus, by a single impulse, upwards of thirty feet, and to dart through the water with the fleetness of a bird upon the wing. The Shrimp frequents the shallows, and congregates in numerous shoals, which leap from the surface, as I have often seen. The capture of them is often left to the children of the fishermen, who, wading in the shoal water, with a net fixed at the end of a pole, take them with much ease.

Under the appellation of Shell-*fish* are familiarly included animals having little connection with each other, and still less with fishes. The Fish, the Crab, and the Oyster belong, in fact, to three of the grand sections into which the animal kingdom is distributed; and though the last two agree in being invested with what is, in common parlance, called "a shell," yet the crust of the one bears no analogy in form, structure, or composition to the shell of the other. Again: those animals which, like the Oyster, are covered with true calcareous shells, differ greatly from each other: some, as the Periwinkle and the Whelk, being animals of much higher grade in the scale of development than others, as the Oyster or Scallop. The former crawl with ease on a broad fleshy disk, as we have all seen in the case of the garden Snail, an animal closely allied to them; they have a distinct head, with tentacles, jaws, and often with eyes; but the latter have no power of crawling, being, for the most part, confined to one spot, no head, no eyes, no tentacles, and no jaws, but are shut up within their two shells, which can be opened

only to a small extent during the life of the animal.
Yet we must not for a moment suppose that these
creatures are unhappy, or that the meanest occupant
even of a bivalve shell is not supplied with everything that could conduce to its welfare. It is SIN
alone that is the source of unhappiness. I will just
point out one or two particulars in which the Divine
care for these creatures is manifest. All of them
have the vital parts of the body protected by a thick
fleshy coat, somewhat projecting at the edges, called
the mantle: the surface of this organ has the power
of forming the shell, by depositing stony matter in
a sort of glutinous cement, which soon hardens into
a thin layer of shell. If a little piece were broken off
the edge of the shell of a Whelk, when alive, the
animal would press the surface of the mantle against
the fracture, and pass it several times over the place;
a very thin transparent film would then be seen to
fill up the space, which in the same way it would
increase in thickness, until in a few days we could
scarcely distinguish the renewed part from the
other, or tell that the shell had been broken, except,
perhaps, by a slight variation in colour. As the animal grows, it wants a larger shell; and the mantle
affords the means of increasing its size: the front
edge of this organ is thicker than the rest, and is
called the collar; and it is by thrusting this round
the edge of the shell, while stony matter is poured
out from its surface, that an addition is made to it.
In the Bivalves, or those whose shells open and shut
like the covers of a book, as the Oyster, the mantle
is twofold, covering the body on each side, just within

each shell. Instead of a collar, each leaf of the mantle is here fringed with a series of delicate fleshy threads, which secrete the exterior part of the shell, by being thrust out round the edge; while the whole surface of the mantle deposits the beautiful, rainbow-tinted, pearly substance with which the interior is coated.

Instead of the fleshy belly on which the Univalves glide along, the Bivalves are furnished with a peculiar organ, which in some species serves the purpose of motion. The Oyster, however, and some other species, have no power of changing their position, but are, as it were, cemented to the rock on which the spawn first chanced to fall. The Mussel, again, is fastened, but in a different manner, being moored by a cable of silken threads, which it spins from its own body. But the Cockle, which is eaten by the poor on many of our shores, is enabled to move with considerable rapidity by means of the organ to which I have just alluded. It is somewhat like a tongue, and can assume a great variety of shapes. The Cockle burrows in the mud: having lengthened and stiffened its tongue or foot, it pushes it as far as it can reach into the mud; then bending the tip into a hook, it forcibly contracts it, and thus brings its body, shell and all, into the hole. The Razor-shell, a shell common on sandy beaches, of a long narrow form, has this power still more remarkably developed.

Many of the islands which stud the sea around the north and west coasts of Scotland are remarkable for the stern grandeur of their precipitous cliffs. One might almost imagine that the surges of the mighty

Atlantic, dashing against them for ages with unbroken fury, had undermined their solid foundations,
and worn for themselves numerous passages, leaving
only columnar rocks of vast height, detatched from
one another, though of similar formation and construction. Such a rock is the Holm of Noss, apparently severed from the Isle of Noss, from which it
is about a hundred feet distant; but the cliffs are
of stupendous height, and far below, in the narrow
gorge, the raging sea boils and foams, so that the
beholder can scarcely look downward without horror.
But stern necessity impels men to enterprises, from
which the boldest would otherwise shrink: to obtain
a scanty supply of coarse food for himself and family,
the hardy inhabitant of the Orkneys dares even the
terrors of the Holm of Noss. In a small boat, with
a companion or two, he seeks the base of the cliff;
and leaving them below, he fearlessly climbs the precipice, and gains the summit. A thin stratum of
earth is found on the top, into which he drives some
strong stakes; and having descended and performed
the same operation on the opposite cliff, he stretches
a rope from one to the other, and tightly fastens it.
On this rope a sort of basket, called a cradle, is
made to traverse, and the adventurous islander now
commits himself to the frail car, and suspended
between sea and sky, hauls himself backward and
forward by means of a line. And do you ask what
prize can tempt man to incur such fearful hazard,
lavish of his life? It is the eggs and young of a sea-
bird, the fishy taste and oily smell of whose flesh
would present little gratification to any whose senses

were not made obtuse by necessity. The Gannets and Guillemots dwell in countless myriads on these naked rocks, laying their eggs and rearing their progeny wherever the surface presents a ledge sufficiently broad to hold them. Their immense numbers render them an object of importance to the inhabitants of these barren islands, who derive from them, either in a fresh state or salted and dried, a considerable portion of their sustenance.

In some other situations the fowlers have recourse to a still more hazardous mode of procedure. The cliffs are sometimes twelve hundred feet in height, and fearfully overhanging. If it is determined to proceed from above, the adventurer prepares a rope, made either of straw or of hog's bristles, because these materials are less liable to be cut through by the sharp edge of the rock. Having fastened the end of the rope round his body, he is lowered down by a few comrades at the top to the depth of five or six hundred feet. He carries a large bag affixed to his waist, and a pole in his hand, and wears on his head a thick cap, as a protection against the fragments of rock which the friction of the rope perpetually loosens; large masses, however, occasionally fall and dash him to pieces.

Having arrived at the region of birds, he proceeds with the utmost coolness and address; placing his feet against a ledge, he will occasionally dart many fathoms into the air, to obtain a better view of the crannies in which the birds are nestling, take in all the details at a glance, and again shoot into their haunts. He takes only the eggs

and young, the old birds being too tough to be
eaten. Caverns often occur in the perpendicular
face of the rock, which are favourite resorts of the
fowls; but the only access to such situations is by
disengaging himself from the rope, and either holding the end in his hand, while he collects his booty,
or fastening it round some projecting corner. I

FOWLING IN ORKNEY.

have heard of an individual, who, either from choice
or necessity, was accustomed to go alone on these
expeditions; supplying the want of confederates
above by firmly planting a stout iron bar in the

earth, from which he lowered himself. One day, having found such a cavern as I have mentioned, he imprudently disengaged the rope from his body, and entered the cave with the end of it in his hand. In the eagerness of collecting, however, he slipped his hold of the rope, which immediately swung out several yards beyond his reach. The poor man was struck with horror; no soul was within hearing, nor was it possible to make his voice heard in such a position; the edge of the cliff so projected that he never could be seen from the top, even if any one were to look for him; death seemed inevitable, and he felt the hopelessness of his situation. He remained many hours in a state bordering on stupefaction; at length he resolved to make one effort, which, if unsuccessful, must be fatal. Having commended himself to God, he rushed to the margin of the cave and sprang into the air, providentially succeeded in grasping the pendulous rope, and was saved.

Sometimes it is thought preferable to make the attempt from below: in this case, several approach the base in a boat; and the most dexterous, bearing a line attached to his body, essays to climb, assisted by his comrades, who push him from below with a pole. When he has gained a place where he can stand firmly, he draws up another with his rope, and then another, until all are up, except one left to manage the boat. They then proceed in exactly the same manner to gain a higher stage, the first climbing and then drawing up the others: and thus they ascend till they arrive at the level of the birds,

K

when they collect and throw down their booty to the
boat. Sometimes the party remains several days on
the expedition, sleeping in the crannies and caverns.
This mode is attended with peculiar hazard; for as
a man often hangs suspended merely from the hands
of a single comrade; it occasionally happens that the
latter cannot sustain his weight, and thus lets him
fall, or is himself drawn over the rock, and shares in
his companion's miserable death.

GUILLEMOT AND GANNET.

The object of these daring adventures, which bring
to mind the words of Shakspeare,

"Half way down
Hangs one that gathers samphire—dreadful trade!"

Is chiefly the Guillemot (*Uria Troile*), a bird some-

what like the Penguin, but with a pointed beak.
The Gannet (*Sula Bassana*) is of the Pelican tribe,
and is confined, at least in large congregations, to
one or two localities: of which the principal are the
Bass Rock on the east coast of Scotland, and St

THE BASS ROCK.

Kilda, the most western of the Hebrides. On these
rocky isles they assemble in such countless hosts
that they can only be compared to a swarm of bees,
or to a shower of snow, the air being filled with
them. The inhabitants of the latter isle are said

to consume twenty-two thousand of the young birds every year, besides eggs. They are powerful birds upon the wing, and pursue with much eagerness the shoals of herrings and pilchards, on which they pounce with the perpendicular descent of a stone. Buchanan conjectures that the Gannets destroy more than one hundred millions of herrings annually. In flying over Penzance some years since, a Gannet's attention was arrested by a fish lying on a board. According to custom, down he swooped on the prey; but his imprudence cost him his life; and it was found that from the impetus of his descent, the bill had quite transfixed the board, though an inch and a quarter in thickness. The fishermen take advantage of this habit, to allure the bird to its destruction; for they fix a fresh herring to a board, and draw it after a sailing boat with some rapidity through the waves; by which many are killed in the manner just narrated. The apparatus by which this bird is furnished for its aërial powers, as well as for aiding its arrowy descent, is very beautiful and instructive. Professor Owen, by inserting a tube into the windpipe, was enabled to inflate the whole body with air, and found that air-cells communicating with each other, pervaded every part, separating even the muscles from each other, and isolating the very vessels and nerves; and penetrating the bones of the wing. A large air-cell was found to be placed in front of the forked-bone, or clavicles, which was furnished with muscles, whose action was instantaneously to expel the air, and thus in a moment to deprive the bird of that buoyancy, so necessary for its flight, but equally detrimental to its swoop.

In some interesting observations, by Colonel Montagu, on the habits of this bird in captivity, the same fact is noticed. When the bird was placed on the water of a pond, nothing could induce him to attempt to dive, and from the manner of his putting the bill, and sometimes the whole head, under water, as if searching for fish, it appears that the prey is frequently so taken. It is probable more fish are caught in their congregated migrations, when the shoals are near the surface, than by their descent upon wing; for the herrings, pilchards, mackerel, and other gregarious fishes, cannot at that time avoid their enemy, who is floating in the midst of profusion. In the act of respiration there appears to be always some air propelled between the skin and the body of this bird, as a visible expansion and contraction is observed about the breast; and this singular conformation makes the bird so buoyant that it floats high on the water, and does not sink beneath the surface, as observed in the cormorant and shag. The legs are not placed so far behind as in such of the feathered tribe as procure their subsistence by immersion; the Gannet, consequently, has the centre of gravity placed more forward; and when standing, the body is nearly horizontal, like a goose, and not erect like a cormorant.

The Gannet collects a slight heap of withered grass and dry sea-weeds, on which it lays and hatches its eggs. They perform this duty by turns, one foraging while the other sits. The roamer, after a predatory excursion, returns to his partner, with five or six herrings in his gorge; these she very complacently pulls out one by one, with much address. Marten says that they frequently rob each other, and that one which had pillaged a nest, artfully flew out towards the sea with the spoil, and returned again, as if it had gathered the stuff from a different quarter. .The owner, though at a distance from his nest, had observed the robbery, and waited the return of the thief, which he attacked with the utmost fury. "This bloody battle," adds the narrator, "was fought above our heads, and proved fatal to the thief, who fell dead so near our boat, that our men took him up, and presently dressed and ate him."

THE ARCTIC SEAS.

PERHAPS in few respects is the character of modern times contrasted with that of antiquity in a
higher degree, than in that enterprising spirit which
prompts men to penetrate distant regions, submit-
ting to unheard-of privations, and braving new difficulties and dangers, not only from the stimulus of
expected gain, but often from the mere love of
knowledge, a desire of gratifying that insatiable and
laudable curiosity, in which all science has its origin.
The ancient nations, bold and intelligent as they
were, knew little of geographical research: precluded from venturing to the north by the dread of
frost, and to the south by the scorching heat of the
sun, both of which their fears so magnified that they
deemed it physically impossible for man to exist in
either the one or the other; their expeditions, in
peace and war, seem to have been well-nigh bounded
by the temperate zone. Thus it happened, that up
to the fifteenth century hardly a fourth of the habit-
able globe was known to the polished nations of
Europe. But then a new era commenced: the discovery of one important law, that the magnetized
needle points always northward, gave a precision to
navigation, and inspired a degree of confidence in
the mariner, which soon led to highly interesting
and unexpected results. The torrid zone was tra-

versed; that terrible "Cape of Storms,"* the south
ern point of Africa, was doubled; a new world was
discovered in the western hemisphere; and commercial enteprise led the hardy sons of western Europe
to dare even the icy horrors of the Poles. Of these
the Biscayans seem to have been the first, for we
find them engaged in the northern whale fishery as
early as the year 1575. Before the end of the six-
teenth century, the English had engaged in the same
enterprise, fishing first on the coast of North America, and after a while in the vicinity of Spitzbergen.
The Dutch soon followed, and other nations were not
slow in prosecuting the same lucrative employment.

Nature in these regions wears an aspect of awful
majesty and grandeur, unrelieved by the softer and
gentler beauties which distinguish her in the south.
In the islands of these seas no meadows smile
in emerald verdure, no waving corn-fields gladden
the heart of man with their golden undulations;
no songs of jocund birds usher in the morning,
nor is the evening soothed with the indefinable
murmur of myriads of humming insects. All is
dreary solitude; and the death-like silence that
pervades the scene, inspires a feeling of involuntary awe, as if the hardy explorer had intruded
into a region where he ought not to be. The
most northern land known to exist is that of the
islands of Spitzbergen, the extreme point of which
approaches to within ten degrees of the Pole. The

* This was the name given to the extreme point of Africa by its dis-
coverer, Bartholomew Diaz: but, on his return to Portugal, King John
II. considered the discovery so auspicious, that he changed the name to
"The Cape of Good Hope," which it still retains.

coast is generally lofty and precipitous, and is visible in clear weather at a great distance, presenting the peculiar features of Arctic scenery in great perfection. The rocks rise in bold and naked grandeur, their summits shooting into innumerable peaks and ridges, and needles, of fantastic forms, reminding the beholder of the domes and spires of a vast city. Most of these are of dark colours, standing out in bold relief against the sky; but their appearance is rendered highly picturesque by the vivid contrasts continually presented by the broad patches of unsullied snow capping their summits, or resting on the ledges and terraces into which their surface is broken, as well as by the glistening accumulations of ice, which fill the valleys nearly to the level of the mountain tops. In approaching the coast in summer, the view is often concealed by the dense fogs so prevalent in that season: suddenly the mist disperses, and these broad contrasts, shown out in startling distinctness beneath a cloudless sun, seem like the sudden creation of a magician's wand. The well-defined outline, and sharp edge of the hues of the picturesque scenery, render it perfectly distinct at a distance at which, in a more southern clime, land would present but a dim and shadowy haze. The objects described may often be clearly seen and well distinguished at the distance of forty miles; and if, after sailing towards the land for four or five hours before a smart breeze, the atmosphere should become slightly charged with mist, the scene might be apparently even more distant than at first. Thus a phenomenon, reported by one of the earlier

Danish navigators, which caused no little astonishment, may be readily accounted for. He had made the eastern coast of Greenland, and had been sailing towards it for many hours with a fair wind; but seeing that the land seemed to be no nearer, he became alarmed, and immediately shifted his course back to Denmark, attributing the failure of his voyage to the influence of loadstone rocks, hidden beneath the sea, which arrested the progress of his vessel.

The peculiar stratification of the rocks in these regions often causes them to assume a walled or castellated appearance, the angles being as sharp and clean as if cut with a mason's tool. Some of their forms resemble so strongly the works of art, that one can scarcely believe them to be freaks of nature. A magnificent instance of such regularity occurs on the coast of Spitzbergen. Near the head of King's Bay, there are seen, far inland, three piles of rock of regular shape, well known to the whalers by the appellation of the Three Crowns. "They rest on the top of the ordinary mountains, each commencing with a square table, or horizontal stratum of rock, on the top of which is another, of similar form and height, but of a smaller area; this is continued by a third, and a fourth, and so on, each succeeding stratum being less than the next below it, until it forms a pyramid of steps, almost as regular to appearance as if worked by art."*

The most prominent object in these dreary seas is ice. Even on the land, a large portion of the ground is concealed by perpetually-accumulating ice, while the same substance covers to a great extent the sur-

* Scoresby.

face of the ocean. There is scarcely a more beautiful object than one of the towering icebergs that so abound in these regions, and that annually come down upon the southern current, into the temperate zone. I have seen numbers of these floating islands, of dazzling whiteness, on the coast of Newfoundland, whither they are brought every spring out of Baffin's Bay. They do not long endure their transition, but soon melt away in the warm waters of the Atlantic, though they are sometimes seen on the coast of the United States, as far down as Philadelphia. In watching some small ice-islands, which, having drifted into the ports of Newfoundland, have grounded in shoal water, I have been surprised to observe how very rapid is their dissolution, even in the month of April. Some large ones, however, are frequently seen in the bays of that country, even in July. They are often of vast dimensions: one seen by Ross, in Baffin's Bay, was estimated to be nearly two miles and a half long, two miles wide, and fifty feet high. Of course this estimate respects only that part which is visible above the surface of the water; but this is a very small portion of its actual bulk. The relative proportion of the part which is exposed to that which is submerged, varies according to the character of the ice: in Newfoundland the part under water is usually considered to be ten times greater than that exposed, but if the ice be porous, it is not more than eight times greater; while, on the other hand, Phipps found that of dense ice, fourteen parts out of fifteen sunk. These floating icebergs are various in form; sometimes rising into pointed spires, like steeples; sometimes taking the form of a conical hill; sometimes that of an overhanging cliff, of most threatening brow.   I have seen some resemble

ICEBERG SEEN IN BAFFIN'S BAY.

the form of a couching lion; but, perhaps, the most ordinary form is that of an irregular mass, higher at one end than at the other.   In the Arctic seas they often present sharp edges and spiry points; but in their progress southward, the gradual influence of climate smooths their unevenness, and gives their surface a rounded outline.   The action of the waves on the portion beneath the surface, undermining the sides and wearing away the projections, continually alters the position of the centre of gravity; and sometimes the effect of this is to cause the whole gigantic mass to roll over with a thundering crash, making the sea to boil into foam, and causing a swell

that is perceptible for miles. When a boat or even a
ship is in immediate proximity to an iceberg in such
circumstances, the danger is imminent; but if viewed

SWELL AMONG ICE.

from a secure distance, the sight is a very interesting
one. The first iceberg I ever saw, and one of large
size, thus rolled about one-third over while I beheld
it, entirely altering its apparent form. Sometimes
the effect of the wave's action is to cause a large
fragment to fall off, or a crack will extend through
the whole mass with a deafening report, or the entire
iceberg will fall to pieces, and strew the ocean with
the fragments, like the remnants of a wreck. Late
in the summer they often become very brittle, and
then a slight violence is sufficient to rupture them.
Seamen avail themselves of the shelter afforded by

L

ice-islands to moor the ship to them in storms, carrying an anchor upon the ice, and inserting the fluke in a hole made for the purpose. In the state just alluded to, such is the brittleness of the substance, that one blow with an axe is sometimes sufficient to cause the immense mass to rend asunder with fearful noise, one part falling one way, and another in the opposite, often swallowing up the ill-fated mariner, and crushing the gallant bark.

SHIP BESET IN ICE.

Contact with floating icebergs, when a ship is under sail, is highly dangerous. From the coolness of the air in their immediate neighbourhood, the moisture of the atmosphere is condensed around them; and hence they are often enveloped in fogs, so as to be invisible within the length of a few fathoms. A momentary relaxation of vigilance on the part of the mariner, may bring the ship's bows on the submerged part of an iceberg, whose sharp needle-like points, hard as rock, instantly pierce the planking, and perhaps open a fatal leak. Many lamentable shipwrecks have resulted from this cause. In the long heavy swell, so common in the open sea, the peril of floating ice is greatly increased, as the huge angular masses are rolled and ground against each other with a force that nothing can resist.

These ice-islands are quite distinct in their nature from the field-ice, which so largely overspreads the surface of the sea, and are believed to be entirely of land formation, consisting of fresh water frozen. The process of their formation is interesting: the glens and valleys in the islands of Spitzbergen are filled up with solid ice, which has been accumulating for uncounted ages; these are the sources from whence the floating icebergs are supplied. Perhaps as long ago as the creation of man, or at least as the deluge, these glaciers began in the snows of winter; the summer sun melted the surface of this snow, and the water thus produced, sinking down into that which remained, saturated it and increased its density. The ensuing winter froze this into a mass of porous ice, and superadded a fresh surface of snow. The same process again going on in summer, of water percolating through the porous crystals, which in its

turn was refrozen, soon changed the lowest stratum
into a mass of dense and transparent ice. Centuries
of alternate winters and summers have thus produced
aggregations of enormous bulk. Scoresby mentions
one of eleven miles in length, and four hundred feet
in height at the seaward edge, whence it slopes upward and backward till it attains the height of six-
teen hundred feet; an inclined plane of smooth
unsullied snow, the beauty and magnitude of which
render it a very conspicuous landmark on that inhospitable shore. The upper surface of a land iceberg is
usually somewhat hollow, and during the summer the
concavities are filled with pools or lakes of the purest
water, which often wears channels for itself through
the substance, or is precipitated in the form of a
cataract over the edge. The water freezing in fissures
thus produced, and expanding with irresistible force,
tears off large fragments from the outer edge, which
are precipitated into the ocean; and high spring
tides, lashed by storms, undermine portions of the
base, and produce the same effect. The masses thus
dislodged float away, and form ice-islands. When
newly broken, the fracture is said to present a
glistening surface of a clear greenish blue, approaching an emerald green; but of such as I have myself
had an opportunity of examining in Newfoundland,
the hollows were of the purest azure.

"On an excursion to one of the Seven Icebergs,"
says Mr. Scoresby, "in July, 1818, I was particularly fortunate in witnessing one of the grandest
effects which these polar glaciers ever present. A
strong north-westerly swell having for some hours

been beating on the shore, had loosened a number of fragments attached to the iceberg, and various heaps of broken ice denoted recent shoots of the seaward edge. As we rode towards it, with a view of proceeding close to its base, I observed a few little pieces fall from the top; and while my eye was fixed upon the place, an immense column, probably fifty feet square, and one hundred and fifty feet high, began to leave the parent ice at the top, and leaning majestically forward, with an accelerated velocity fell with an awful crash into the sea. The water into which it plunged was converted into an appearance of vapour or smoke, like that from a furious cannonading. The noise was equal to that of thunder, which it nearly resembled. The column which fell was nearly square, and in magnitude resembled a church. It broke into thousands of pieces. This circumstance was a happy caution, for we might inadvertently have gone to the very base of the icy cliff, from whence masses of considerable magnitude were continually breaking."*

" 'Tis sunset: to the firmament serene
The Atlantic wave reflects a gorgeous scene;
Broad in the cloudless west, a belt of gold
Girds the blue hemisphere; above unroll'd,
The keen, clear air grows palpable to sight,
Embodied in a flush of crimson light,
Through which the evening star, with milder gleam,
Descends to meet her image in the stream.
Far in the east, what spectacle unknown
Allures the eye to gaze on it alone?
—Amidst black rocks, that lift on either hand
Their countless peaks, and mark receding land;

* Arctic Regions, i. 104.

Amidst a tortuous labyrinth of seas
That shine around the arctic Cyclades;
Amidst a coast of dreariest continent,
In many a shapeless promontory rent;
—O'er rocks, seas, islands, promontories spread,
The Ice-Blink rears its undulated head;
On which the sun, beyond th' horizon shrined,
Hath left his richest garniture behind;
Piled on a hundred arches, ridge by ridge,
O'er fixed and fluid strides the Alpine bridge,
Whose blocks of sapphire seem to mortal eye
Hewn from cerulean quarries of the sky;
With glacier battlements, that crowd the spheres,
The slow creation of six thousand years,
Amidst immensity in towers sublime,
Winter's eternal palace, built by Time.
All human structures by his touch are borne
Down to the dust; mountains themselves are worn
With his light footstep; *here* forever grows,
Amid the region of unmelting snows,
A monument; where every flake that falls
Gives adamantine firmness to the walls.
The sun beholds no mirror, in his race,
That shows a brighter image of his face;
The stars, in their nocturnal vigils, rest
Like signal fires on its illumined crest;
The gliding moon around the ramparts wheels,
And all its magic lights and shades reveals;
Beneath, the tide with idle fury raves
To undermine it through a thousand caves,
Rent from its roof though thundering fragments oft
Plunge to the gulf, immovable aloft,
From age to age, in air, o'er sea, on land,
Its turrets heighten, and its piers expand."*

By far the greatest portion of the ice met with in navigating these seas is of marine formation. During the greater part of the year, in high lati-

* Montgomery's "Greenland," p. 61.

tudes, the process of congelation is always going
on at the surface of the sea. If the wind is high,
the crystals cannot readily unite into a solid form,
but form a spongy mass, called sludge: when this
has become somewhat thick, however, the wind can
no longer act upon the water, so as to raise little
ripples upon it, and the sludge now begins "to
catch;" but the swell prevents one uniform surface
being yet formed, and the consequence is, that small
rounded plates of ice are produced, called "pan-
cakes," the edges of which are raised slightly, by
the constant pressure of one against another. The
cakes in the centre of the freezing mass now begin
to adhere to each other, and thus a solid surface
is produced, which gradually extends both its diameter and its depth. The individual pieces of
which such ice is composed are distinctly to be
traced, even when perfectly consolidated, and present an appearance resembling pavement. But in
calm weather, a thin pellicle of ice is simultaneously produced over the whole surface of the
sea, and the formation of the ice-field is much
more direct and obvious. Single fields have been
seen many leagues in length, and occupying an
area of several hundred square miles; being at
the same time from three to six feet high, and
from ten to twenty deep. The waves produced
by storms break up these fields into smaller pieces,
called floes, and driving one against another with
violence, the edge of one is often lifted upon the
other by the force of the pressure, and *hummocks*
or hills, of various shapes and sizes, are raised upon

them. Ice-fields often acquire a rotatory motion;
and when we consider the immense weight of these
ponderous masses, we shall have an idea of the
irresistible impetus communicated by such a body
in motion. Scoresby calculates one mentioned by
him at ten thousand millions of tons: no wonder,
that coming in contact with a vessel, her iron knees
and oaken timbers should be crushed like a walnut,
or that she should be lifted clean out of the water by
the pressure, and placed high and dry upon the ice!
From this cause arise many of the accidents which
give to the navigation of the Arctic sea its peculiarly-
hazardous character.

When the temperature of the atmosphere is about
two or three degrees above the freezing-point, a
surface of ice, if placed in a horizontal plane, will
melt, not by a general dissolution of its substance,
but so as to leave a multitude of perpendicular
columns, or needles. In the late attempt to reach
the North Pole by boats hauled over the ice, Captain Parry found ice in this condition productive of
no little inconvenience. At the very commencement
of the journey we find it thus noticed:—"June
26.—A great deal of the ice over which we passed
to-day presented a very curious appearance and
structure, being composed, on its upper surface,
of numberless irregular, needle-like crystals, placed
vertically, and nearly close together; their length
varying, in different pieces of ice, from five to ten
inches, and their breadth in the middle about half
an inch, but pointed at both ends. The upper surface of ice having this structure, sometimes looks

like greenish velvet; a vertical section of it, which frequently occurs at the margin of floes, resembles, while it remains compact, the most beautiful satin spar; and asbestos, when falling to pieces. At this early part of the season, this kind of ice afforded pretty firm footing; but as the summer advanced the needles became more loose and movable, rendering it extremely fatiguing to walk over them, besides cutting our boots and feet—on which account the men called them penknives."* The Captain attributes this peculiar structure to the heavy drops of rain piercing their way downwards through the ice, and separating it into needles.

There is no phenomenon that more forcibly brings before the mind of a stranger the novelty of his position, than the absence, on entering within the Arctic Circle, of that constant alternation of day and night, which we are accustomed to consider as inseparable from the constitution of our world. We have learned this fact in our elementary treatises on Geography, but yet it is difficult to realise to the mind a perpetual day, an unsetting sun. When the sun's disk is obscured by a fog, it is no uncommon thing for sailors to ask each other if it be night or day: and Phipps, on his return voyage, thought the sight of a star an occurrence of sufficient moment to be inserted in his journal. "August 24th. —We saw Jupiter: the sight of a star was now become almost as extraordinary a phenomenon as the sun at midnight, when we first got within the Arctic Circle." Our voyagers usually seek the

* Narrative of an Attempt, &c., p. 61.

9

Arctic Ocean in spring, and leave it at the approach of autumn; a winter residence there being dreaded as one of the direst calamities that can befall them; and therefore, until lately, our knowledge of winter phenomena was very meagre, and mainly derived from the reports of a few unhappy men, by accident compelled to remain in a clime so inhospitable. By the experience of the officers and crews engaged in the recent voyages of discovery, we have become nearly as familiar with the phenomena of the long winter's night, as with those of the short summer's day. In Spitzbergen the day is rather more than four months long: the night is of the same duration, and in the two months which intervene between the sun's constant presence and his constant absence, that luminary rises and sets as with us. But the appearance of the sun in spring is accelerated, and its disappearance in autumn retarded, a few days, by the influence of refraction; so that it is actually seen somewhat longer than it is invisible. Thus Captain Parry, at Melville Island, saw the sun on the first of February, which was about four days earlier than its actual elevation above the horizon; in like manner it remained visible until the 11th of November, whereas it had actually sunk beneath the horizon on the 7th. Then the darkness of the Arctic winter is not total and incessant; even in the depth of the season, at Spitzbergen, there is a faint twilight for six hours each day, and this is longer and brighter in proportion to the distance from midwinter on either hand. The moon also shines in

those clear skies with peculiar brilliance, and is
often visible twelve or fourteen days without setting. There is, moreover, a large proportion of
the time, in which the Aurora Borealis illumines

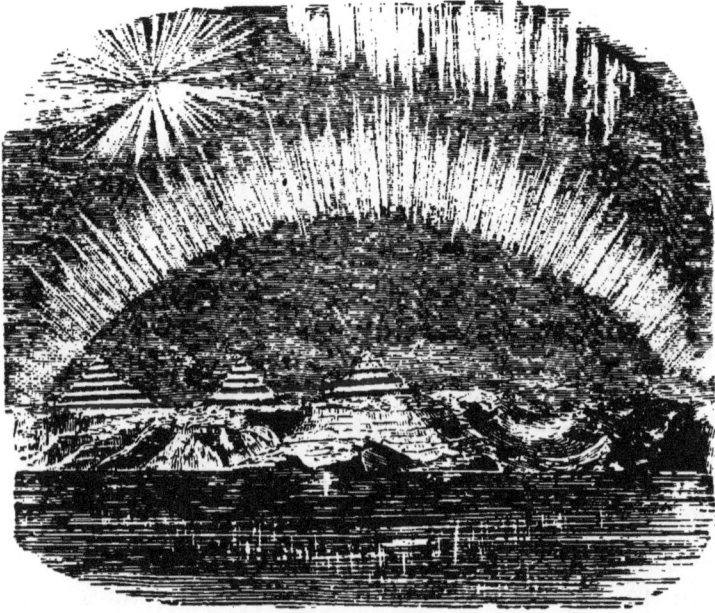

AURORA BOREALIS.

The scene is in the vicinity of the Three Crowns on the Coast of
Spitzbergen. See p. 106.

the heavens, and sometimes with an intensity little
inferior to moonlight. This interesting meteor is
occasionally seen in England, but very rarely with
that brilliance with which it shines in the Frigid
Zone, and in the northern parts of America. In
Newfoundland and Canada I have seen many specimens of the Aurora, and some splendidly coloured
with blue, green, and red hues; sometimes the

whole sky has been flushed with intense crimson,
which, reflected from the snow beneath, had an
awful, though beautiful appearance. The following details of one which I observed in Lower Canada, in February, 1837, will give a notion of the
appearance of this meteor in its more usual state.
"I first observed it about half-past eight o'clock:
a long, low, irregular arch of bright yellow light
extended from the north-east to the north-west,
the lower edge of which was well defined; the sky
beneath this arch was clear, and appeared black, but
it was only by contrast with the light, for on examination, I could not find that it was really darker
than the other parts of the clear sky. The upper
edge of the arch was not defined, shooting out rays
of light towards the zenith: one or two points in
the arch were very brilliant, which were varying in
their position. Over head, and towards the south,
east, and west, flashings of light were darting from
side to side: sometimes the sky was dark, then
instantly lighted up with these fitful flashes, vanishing and changing as rapidly; sometimes a kind
of crown would form around a point south of the
zenith, consisting of short converging pencils. At
nine o'clock, the upper and southern sky was filled
with clouds or undefined patches of light, nearly
stationary; the eastern part, near the top, being
bright crimson, which speedily spread over the upper
part of the northern sky. A series of long converging pencils was now arranged around a blank space
about 15° south of the zenith, the northern and
eastern rays blood-red, the southern and western

pale yellow; the redness would flash about, as did
the white light before, still not breaking the general
form of the corona. In a few minutes all the red
hue had vanished, leaving the upper sky nearly un-
occupied. The arch also was now totally gone, and
in its place there were only irregular patches of
yellow light, of varying radiance. At a quarter
past nine the upper sky was again filled with pale
flashes: in the north were perpendicular pillars of
light, comparatively stationary. At half-past nine
there was no material change, and at ten all had
assumed a very ordinary appearance, merely large
clouds of pale light being visible."* The cause
which produces these beautiful coruscations of light
in high latitudes has not yet been satisfactorily
known: it seems pretty certain that their origin is
in general far above our atmosphere.

Montgomery alludes to the Aurora in the following beautiful lines:—

> "Midnight hath told his hour: the moon, yet young,
> Hangs, in the argent west, her bow unstrung;
> Larger and fairer, as her lustre fades,
> Sparkle the stars amidst the deepening shades:
> Jewels more rich than night's regalia gem
> The distant Ice-Blink's spangled diadem;
> Like a new morn from orient darkness, there
> Phosphoric splendours kindle in mid-air,
> As though from heaven's self-opening portals came
> Legions of spirits in an orb of flame,—
> Flame that from every point an arrow sends,
> Far as the concave firmament extends:
> Spun with the tissue of a million lines,
> Glistening like gossamer the welkin shines:

* Canadian Naturalist, p. 47.

M

The constellations in their pride look pale
Through the quick trembling brilliance of that veil
Then suddenly converged, the meteors rush
O'er the wide south; one deep vermilion blush
O'erspreads Orion glaring on the flood,
And rabid Sirius foams through fire and blood;
Again the circuit of the pole they range,
Motion and figure every moment change,
Through all the colours of the rainbow run,
Or blaze like wrecks of a dissolving sun;
Wide ether burns with glory, conflict, flight,
And the glad ocean dances in the light."*

This interesting meteor, occurring with more or
less of splendour in rapid succession, added, moreover, to the universal reflection of what light may
proceed from the heavens by the pure whiteness of
the ice and snow, tends greatly to lessen the darkness
of the long and dreary night, though these causes
cannot diminish the cold. The latter was so intense
during the late expeditions of discovery, that the
temperature was 55° below zero, or eighty-seven
degrees below the freezing-point.

The remarkable appearances called mock suns, or
parhelia, are extremely frequent within the Arctic
Circle. Their usual appearance may be thus described. When the sun is not far from the horizon,
one or more luminous circles, or halos, surround it
at a considerable distance; two beams of light go
across the innermost circle, passing through the
centre of the sun, the one horizontally, the other
perpendicularly, so as to form a cross: where these
beams touch the circle, the light is, as it were, concentrated in a bright spot, sometimes scarcely inferior in brilliance to the sun itself; at the corre-

* " Greenland," p. 64.

sponding points in the outermost circle, segments of
other circles, wholly external, come into contact with
it. It is not often that this meteor is seen in the
perfection described: occasionally the circles are too

MOCK SUNS.

The scene is the coast of Barrow's Strait.

faint to be visible; and the mock suns alone are
seen in the usual places, and sometimes but one or
two of them. Another singular appearance, called
the fog-bow, of great beauty and interest, is thus
described by Mr. Scoresby: "The intense fogs
which prevail in the Polar Seas, at certain seasons,
occasionally rest upon the surface of the water, and
reach only to an inconsiderable height. At such

times, though objects situated on the water can
scarcely be discerned at the distance of a hundred
yards, yet the sun will be visible and effulgent.
Under such circumstances, on the 19th July, 1813,
being at the topmast head, I observed a beautiful
circle of about 30° diameter, with bands of vivid
colours depicted on the fog. The centre of the circle
was in a line drawn from the sun through the point
of vision, until it met the visible vapour in a situation exactly opposite the sun. The lower part of
the circle descended beneath my feet to the side of
the ship; and although it could not be a hundred feet
from the eye, it was perfect, and the colours distinct.
The centre of the coloured circle was distinguished
by my own shadow, the head of which, enveloped
by a halo, was most conspicuously portrayed. The
halo or glory was evidently impressed on the fog, but
the figure appeared to be a shadow on the water, the
different parts of which became obscure in proportion
to their remoteness from the head, so that the lower
extremities were not perceptible. I remained a long
time contemplating the beautiful phenomenon before
me.   Notwithstanding the sun was brilliant and
warm, the fog was uncommonly dense beneath. The
sea and ice, within sixty yards of the ship, could
scarcely be distinguished. The prospect thus circumscribed served to fix the attention more closely
on the only interesting object in sight, whose radiance and harmony of colouring, added to the singular appearance of my own image, were productive of
sensations of admiration and delight."*   I have

* Arct. Reg. i. 394

myself had the pleasure of witnessing this beautiful phenomenon, precisely as described above, and in the same circumstances: it was in the month of August, 1828, on the coast of Newfoundland, and was viewed from the shrouds of a vessel projected on the surface of a dense but shallow fog. Sometimes there are several coloured circles surrounding each other, with a common centre.

The cause of these appearances seems to be the unequal refraction of the rays of light by passing through media of varying density. To a similar origin may be ascribed those distortions and repetitions of objects near the horizon, called *looming*, which are occasionally witnessed even in this country, but in the northern seas are very frequent and amusingly fantastic. The ice around the horizon, either almost flat or varied only by slight irregularities of surface, will appear raised into a lofty wall, and the irregularities elevated into numberless spires or towers or pinnacles. Ships will have their hulls magnified into castles; or the hull will be diminished to a narrow line, and the masts and sails drawn up to a ridiculous length; or some of the sails will be unduly elevated, while others are as unnaturally flattened. But more singular than this is the frequent repetition of the object in the sky just above it. Thus above the spired and turreted wall of ice will be seen on the sky another wall exactly corresponding to it, but upside-down; spire meeting spire, and tower tower. Above a ship will be an inverted figure of the same ship, as palpable and apparently as real as the true one. This I once saw, in two vessels in the Gulf of St.

Lawrence. Sometimes another image may be seen above the inverted one, and sometimes, but very rarely, even a fourth. In such cases, the third is always in a right position, and the fourth inverted like

DISTORTIONS OF IRREGULAR REFRACTION.

the second. An image of a vessel is sometimes seen projected upon the sky, when nothing corresponding to it is visible below, the real object being far below the horizon. Mr. Scoresby thus saw his father's ship, the Fame, drawn upon the sky, and by the aid of a telescope could make her out so distinctly as to pronounce with confidence upon her identity, when, by comparing notes afterwards, it was found that she was thirty miles distant at the time, and seventeen miles from the extreme point of vision. Somewhat allied to this is the bright gleam seen by night above field-ice, called *ice-blink*, which is often very serviceable in indicating the presence of ice below the horizon; or by the dark spots and patches in it corresponding to the openings of water, directing the seamen, when beset, how to reach them, when otherwise their existence would be unknown.

The officers engaged in the late expeditions of discovery have remarked the impossibility of correctly measuring distances by the eye when traversing a plain of unbroken snow or ice. Sometimes in travelling, they would discern what appeared to be a rock or a hummock of ice of considerable magnitude, and at a great distance; and having set their course by it, rejoicing that for some time the painful straining of the sight in keeping the direction would be spared by the advantage of so conspicuous a mark, in a minute or two they would reach it, when it would turn out to be some insignificant object, scarcely larger than a hat.

Some of the effects of intense cold, as witnessed in these northern climes, are mentioned by Mr. Scoresby, and are interesting, because they never occur in our own country. After mentioning a very sudden depression of the temperature, he says:—"This remarkable change was attended with singular effects. The circulation of the blood was accelerated; a sense of parched dryness was excited in the nose; the mouth, or rather the lips, were contracted in all their dimensions, as by a sphincter, and the articulation of many words was rendered difficult and imperfect; indeed, every part of the body was more or less stimulated or disordered by the severity of the cold. A piece of metal, when applied to the tongue, instantly adhered to it, and could not be removed without its retaining a portion of the skin; iron became brittle, and such as was at all of inferior quality, might be fractured by a blow; brandy of English manufacture and wholesale strength was frozen; quicksilver, by a single process, might have been consolidated; the sea, in some places, was in the act of freezing, and in others appeared to smoke, and produced, in the formation of *frost-rime*, an obscurity greater than that of the thickest fog. The subtle principle of magnetism seemed to be, in some way or other, influenced by the frost; for the deck-compasses became sluggish, or even motionless, while a cabin-compass traversed with celerity. The ship became enveloped in ice; the bows, sides, and lower rigging were loaded; and the rudder, if not repeatedly freed, would in a short time have been rendered immovable."* In winter, however, the temperature being much lower, the effects of intense cold are more manifest. Egede observes of Disco Island in the month of January, "The ice and hoar-frost reach through the chimney to the stove's mouth, without being thawed by the fire in the day-time. Over the chimney is an arch of frost, with little holes, through which the smoke discharges itself. The doors and walls are as if they were plastered over with frost, and, which is scarcely credible, beds are often frozen to the bedsteads. The linen is frozen in the drawers. The upper eider-down bed and the pillows are quite stiff with frost an inch thick, from the breath."† Many of these results I have myself

* Arct. Reg. i. 330.   † Crantz, Hist. of Greenland.

witnessed in Newfoundland and Lower Canada, some
of which I have alluded to elsewhere;* in the former
country it is not uncommon for the vapour of a
sleeping-room, condensed on the windows and walls,
to take the form of thin narrow blades of ice standing out horizontally, very closely set together; the
whole making a dense coating, of more than half an
inch in thickness, of spongy frost. In the first winter spent at Melville Island by Captain Parry, an ac-
cumulation of a similar substance was observed, that
was really astonishing. "The Hecla was fitted with
double windows in her stern, the interval between
the two sashes being about two feet; and within
these some curtains of baize had been nailed close in
the early part of the winter. On endeavouring now
to remove the curtains, they were found to be so
strongly cemented to the windows by the frozen
vapour collected between them, that it was necessary to cut them off, in order to open the windows;
and from the space between the double sashes we
removed more than *twelve large buckets full* of ice,
or frozen vapour, which had accumulated in the same
manner."†

The shooting out of crystals of beautiful forms,
when vapour is deposited upon any very cold substance, is a very pleasing phenomenon. The feather-
like hoar-froast, so often seen in winter on stems and
blades of grass, is of this character. But it is in the
icy seas of the north that this beauty is seen in perfection. For an interesting description, we have
again recourse to Mr. Scoresby. "In the course of

* Canadian Naturalist, 350.   † Parry's First Voyage, 146.

the night, the rigging of the ship was most splendidly decorated with a fringe of delicate crystals. The general form of these was that of a feather having half of the vane removed. Near the surface of the ropes was first a small direct line of very white particles, constituting the stem or shaft of the feather; and from each of these fibres, in another plane, proceeded a short delicate range of spiculæ or rays, discoverable only by the help of a microscope, with which the elegant texture and systematic construction of the feather were completed. Many of these crystals, possessing a perfect arrangement of the different parts corresponding with the shaft, vane, and rachis of a feather, were upwards of an inch in length, and three-fourths of an inch in breadth. Some consisted of a single flake or feather; but many of them gave rise to other feathers, which sprang from the surface of the vane at the usual angle. There seemed to be no limit to the magnitude of these feathers, so long as the producing cause continued to operate, until their weight became so great, or the action of the wind so forcible, that they were broken off, and fell in flakes to the deck of the ship."*

In our own winters we are familiar enough with snow; but, probably, few are aware of the exceeding beauty, regularity, and delicacy which mark each individual crystal of this production. In our climate, indeed, the temperature during a fall of snow is rarely low enough for the form of the crystals to be perceived; as they become slightly melted in passing

* Arct. Reg. i. 437.

through the air, and many crystals adhere together, and form the irregular aggregations called *flakes* of snow. The ordinary form is that of a six-rayed star; but the rays are often furnished with minute side rays, like the beards of a feather, or are varied in almost infinite diversity. The angle, however, which is formed in crystalization, is invariably the same, namely, one of 60°; and hence arises their symmetry.

Frost is a powerful antiseptic; as fermentation will not take place in a low temperature, animal substances may be kept without decay for an indefinite period. It is customary for the whalers to take out their meat unsalted, trusting to this well-known quality of cold. Captain Parry's crew, fast locked up in the ice of Melville Island, enjoyed a Christmas dinner of roast beef, perfectly sweet, which had been put on board nine months before. The Mammoth which was dislodged by the falling of a cliff at the mouth of the river Lena, had been preserved from putrefaction for uncounted ages. And more affecting instances of this quality have been witnessed in the bodies of men, who, having died in these icy regions, had lain for years unburied without decay. In 1774, the uncouth form of an apparently-deserted ship was met with, strangely encumbered with ice and snow: on boarding her, a solitary man was found in her cabin, his fingers holding the pen, while before him lay the record which that pen had traced, bearing date twelve years before. No appearance of decay was manifest, save that a little greenish mould had accumulated on his forehead. A strange awe crept over the minds of those who thus first broke in upon his loneliness: for twelve years had that ill-fated bark navigated, through sun and storm, the Polar Sea; and, perhaps, unconsciously solving the problem that had so long baffled human skill and daring, had even crossed the Pole itself.

But it is time that we turn from the consideration of inanimate nature and atmospheric phenomena, to inquire what are the living productions that cheer the loneliness of the Arctic mariner. Of the vegetation of these regions we know little: the dreary level shores of many of the isles are marshy, and densely clothed with various mosses, which, though frozen in winter, revive in the transient summer. The rocks, too, are covered with lichens of various colours; and a few dwarf flowering plants just rise above the thin soil. Nothing like a tree varies the scene, but large trunks of trees are brought, by the currents, from distant regions, and washed upon the sea beach. Some of the Fuci which are common with us are found also on these shores, and doubtless many other species which are unknown to us.

The most notorious of the inhabitants of these dreary seas are the mighty and gigantic Whales. "There is that leviathan, whom thou hast made to play therein." It is in pursuit of these immense creatures, and especially the Greenland species, the "right Whale" of the seamen (*Balæna mysticetus*), that many ships, well-manned and fitted out at great expense, proceed every year from England, Holland, France, and other nations, into the Arctic zone. This valuable animal has produced to Britain 700,000*l.* in

a single year, and one cargo has been known to yield
11,000*l*. It is, therefore, well worth our consideration, and the more particularly, because in its structure and habits there are more than ordinary evidences of that gracious forethought and contrivance,
the tracing of which makes the study of nature so
instructive. The Greenland Whale has no affinity
with fishes; it is as much a mammal as the ox or the
elephant, having warm blood, breathing air, bringing
forth living young, and suckling them with true
milk. It inhabits the Polar Seas, beyond which
there is no satisfactory proof that it has ever been
seen. Its length is from fifty to sixty feet, when
full grown; perhaps, in extremely rare cases, seventy
feet; all statements giving it a greater length than
this, either refer to other species, such as the great
Rorqual, or are gross exaggerations. The form is
rather clumsy, the head being very large, and the
mouth reaching to scarcely less than a fourth of the
total length of the animal. The gullet is so small as
not to admit the passage of a fish so large as a herring; hence its support is derived from creatures of
very small bulk, and apparently insignificant, such
as shrimps, sea slugs, sea blubbers, and animalcules
still smaller, which I will presently notice. But
how does it secure its minute and almost invisible
prey? for without some express provision, these
atoms would be quite lost in the cavity of its
capacious mouth, unless swallowed promiscuously
with the water, which would fill the stomach before a hundredth part of the meal was obtained.
There is a very peculiar contrivance to meet this

exigency; the mouth has no teeth, but from each upper jaw proceed more than three hundred horny plates, set parallel to each other, and very close; they run perpendicularly downwards, are fringed on the inner edge with hair, and diminish in size from the central plate to the first and last, the central one being about twelve feet long. The plates are commonly called *whalebone*, and their substance is well known to everybody; they form an important object of the fishery. The lower jaw is very deep, like a vast spoon, and receives these depending plates, the use of which is this: when the Whale feeds, he swims rapidly just under or at the surface, with his mouth wide open; the water with all its contents rushes into the immense cavity, and filters out at the sides between the plates of the whalebone, which are so close, and so finely fringed, that every particle of solid matter is retained.

Though the Whale, like all other *Mammalia*, is formed for breathing air alone, and is therefore necessitated to come to the surface of the sea at certain intervals, yet those intervals are occasionally of great length. We well know that we could not intermit the process of breathing for a single minute without great inconvenience, and the lapse of only a few minutes would be followed by insensibility and perhaps death. The Whale, however, can remain an hour under water, or, in an emergency, even nearly two hours, though it ordinarily comes up to breathe at intervals of eight or ten minutes, except when feeding, when it is sometimes a quarter of an hour, or twenty minutes submerged. Now the object of breathing is to renew the vital qualities of the blood by presenting it to the air, the oxygen in which uniting with the blood renders it again fit for sustaining life. But if more blood could be oxygenized at once than is wanted for immediate use, and the overplus deposited in a reservoir until wanted, respiration could be dispensed with for a while. This is actually what the wisdom of God has contrived in the Whale. The exhausted blood, which is returned by the veins, having been renewed by its communication with the air in the lungs, is carried to the heart, whence only a part is carried away into the system, the remainder being received into a great irregular reservoir, consisting of a complicated series of arteries, which first lines a large portion of the interior of the chest, then insinuating itself between the ribs, forms a large cushion outside of them near the spine, and also within the spinal tube, and even within the skull. The blood thus reserved is poured into the system as it is needed, and thus prevents the necessity of frequent access to the surface.

It is an object of importance that the act of breathing should be performed with as little effort as possible, and therefore the windpipe is made to terminate not in the mouth, nor in nostrils placed at the extremity of the muzzle. If this were the case it would require a large portion of the head and body to be projected from the water, or else that the animal should throw itself into a perpendicular position; either of which alternatives would be inconvenient when swimming rapidly, as, for example, endeavouring to escape when harpooned. The windpipe, there-

fore, communicates with the air at the very top of
the head, which, by a peculiar rising or hump at that
part, is the very highest part of the animal when
horizontal, so that it can breathe when none of its body
is exposed except the very orifice itself. The Whale
often begins to breathe when a little below the surface, and then the force with which the air is expired
blows up the water lying above in a jet or stream,
which with the condensed moisture of the breath
itself constitutes what are called "the spoutings,"
and which are attended with a rushing noise that may
be heard upwards of a mile. Some naturalists have
maintained that a stream of water is ejected from the
blow-hole in the form of an united column, mounting
high before it falls again in a shower. But from my
own observation on many individuals (seen in the
Atlantic), I incline to the former conclusion; as I
have invariably seen the ejected matter, instead of
forming a column, and falling in a shower, sail away
upon the breeze like a little white cloud. These
were, I suppose, Rorquals: but what is true of one
species, is probably true of all. There are one or
two other beautiful contrivances connected with the
structure of this air-passage, that are well worth no-
ticing. In the agony and terror caused by the blow
of the harpoon, the Whale usually plunges directly
downward into the depths of the sea, and that with
such force that the mouth has been found on returning to the surface, covered with the mud of the bottom; while in some instances the jaws, and in others
the skull, have been fractured by the violence with
which they have struck the ground. A Whale has

been known to descend perpendicularly to the depth
of a mile, as measured by the length of line "run
out;" where the pressure of the immense body of
water above would be equal to a ton upon every
square inch. And Mr. Scoresby mentions a case in
which a boat that was accidentally entangled was
carried down by the Whale, which was presently
captured, and the boat recovered by being drawn up
with the line; but from the intense pressure, the
water had been forced into the pores of the solid
oak, so that it was completely saturated, and sunk
like lead: the paint came off in large sheets, and the
wood thrown aside to be used as fuel, was found to
be useless, for it would not burn. A piece of the
lightest fir-wood, which was in the boat, came up in
exactly the same soaked condition, having totally
lost the power of floating. To resist such a pressure
as this, the blow-holes of the Whale tribe are closed
with a valve-like stopper of great density and elasticity, somewhat resembling India-rubber, which, accurately fitting the orifice, excludes all water from
the windpipe, becoming more tightly inserted in
proportion to the pressure.

But this precaution would be vain, if the structure
of the interior of the mouth were the same as in
other Mammalia. Usually the windpipe and gullet
open into a hollow at the back of the mouth, and
the passage to the nostrils proceeds from it likewise.
The windpipe passes up in front of the gullet, and
the food which passes over the former is prevented
from entering it by a lid or valve, which shuts down
during the act of swallowing, but at other times is

erect. But if such were the construction in the
Whale, the force with which the water rushes into
the mouth would inevitably carry a large portion of
the fluid down upon the lungs, and the animal would
be suffocated. The windpipe is therefore carried
upward in a conical form, with the aperture upon
the top, and this projecting cone is received into the
lower end of the blowing-tube, which tightly grasps
it; and thus the communication between the lungs
and the air is effected by a continuous tube, which
crosses the orifice of the gullet, leaving a space on
each side for the passage of the food.

It is doubtless to give increased power of resistance to the eye of the Whale in the pressure of
enormous depths, that there is a peculiar thickness
in the *sclerotic* coat. This is the part which in man
is usually called the *white* of the eye. "When we
make a section of the whole eye, cutting through the
cornea, the *sclerotic* coat, which is as dense as tanned
leather, increases in thickness towards the back part,
and is full five times the thickness behind that it is
at the anterior part. The fore part of the eye sustains the pressure from without, and requires no additional support; but were the back part to yield,
the globe would then be distended in that direction,
and the whole interior of the eye consequently suffer
derangement. We see, then, the necessity of the
coats being thus remarkably thickened behind."*

Another no less interesting deviation from ordinary
structure is found in the skin; the object still being
defence against external pressure. Every one is pro-

* Paley's Nat. Theol., Bell and Brougham's edit. p. 40.

bably aware that the body of the Whale is encased in
a thick coat of fat, denominated blubber, varying in
diameter from eight inches to nearly two feet in different parts of the animal. It has, however, been only
recently known that this fat lies not under the skin,
but actually in its substance. I shall describe this in
the words of Professor Jacob, who first made known
this interesting peculiarity:—"That structure in
which the oil is deposited, denominated blubber, is
the true skin of the animal, modified certainly for
the purpose of holding this fluid oil, but still being
the true skin. Upon close examination it is found
to consist of an interlacement of fibres, crossing each
other in every direction, as in common skin, but more
open in texture, to leave room for the oil. Taking
the hog as an example of an animal covered with an
external layer of fat, we find that we can raise the
true skin without any difficulty, leaving a thick layer
of cellular membrane, loaded with fat, of the same
nature as that in the other parts of the body; on the
contrary, in the Whale it is altogether impossible to
raise any layer of skin distinct from the rest of the
blubber, however thick it may be; and, in *flensing* a
Whale, the operator removes this blubber or skin
from the muscular parts beneath, merely dividing
with his spade the connecting cellular membrane."*
Such a structure as this, being firm and elastic in the
highest degree, operates like so much India-rubber,
possessing a density and power of resistance which increases with the pressure. But this thick coating of
fat subserves other important uses. An inhabitant

* Dublin Philos. Journ. i. 356.

of seas where the cold is most intense, yet warm-
blooded, and dependent for existence on keeping up
the animal heat, the Whale is furnished in this thick
wrapper with a substance which resists the abstrac-
tion of heat from the body as fast as it is generated,
and thus is kept comfortably warm in the fiercest
polar winters. Again, the oil contained in the cells
of the skin being specifically lighter than water, adds
to the buoyancy of the animal, and thus saves much
muscular exertion in swimming horizontally and in
rising to the surface; the bones, being of a porous or
spongy texture, have a similar influence.

These few particulars in the physiology of these
vast creatures may serve to carry our minds up in
adoring wonder to the mercy as well as wisdom of
the Lord God Almighty, and may give us a glimpse
of the meaning of that glorious truth, "And God
saw everything that He had made, and behold it
was VERY GOOD." Many other instances of beautiful contrivance and design might easily be added,
in the construction of the mouth, the eyes, the fins,
the tail; but all would lead us to the same result:
and these which I have adduced may be taken as
a sample of the rich feast which the study of nature
affords to the Christian student.

The capture of these immense animals, from their
vast strength, the fickle element on which it is pursued, and the horrors peculiar to the Arctic regions,
is an adventure of extraordinary hazard. The ships,
built for the purpose, and strengthened with much
oak and iron, leave the northern parts of this country
early in April, and by the end of the month

usually reach the scene of their enterprise. Arrived within the limits of constant day, an unceasing watch is kept for Whales, by an officer stationed in a snug sort of pulpit, called the *crow's-nest*, made of hoops and canvas, and well secured at the main-topmast head. The boats, which combine strength and lightness, are always kept hanging over the sides and quarters of the ship, ready furnished for pursuit, so that on the appearance of a Whale being announced from aloft, one or more boats can be despatched in less than a minute. Each boat carries a harpooner, whose station is in the bow, a steersman, and several rowers. In an open space in the bow of the boat is placed a line sometimes more than 4000 feet in length, coiled up with beautiful regularity and scrupulous care. The end of this is fastened to the harpoon, a most important weapon, made of the toughest iron, somewhat in the form of an anchor, but brought to an edge and point. Instead of steel being employed, as is commonly supposed, the very softest iron is chosen for this important implement, so that it may be scraped to an edge with a knife. A long staff is affixed to the harpoon, by which it is wielded. The boat is swiftly, but silently, rowed up to the unconscious Whale, and when within a few yards, the harpooner darts his weapon into its body. Smarting and surprised, the animal darts away into the depth of the ocean, but carries the harpoon sticking fast by the barbs, while the coiled line runs out with amazing velocity. A sheeve or pulley is provided, over which it passes; but if by accident it slips out of its place, the friction is so great that

the bow of the boat is speedily enveloped in smoke,
and instances are not unfrequent of the gunwale
even bursting into a flame, or even of the head of the
boat being actually sawn off by the line. To prevent
this, a bucket of water is always kept at hand, to
allay the friction. Accidents even still more tragic
sometimes occur from entanglements of the line.
"A sailor belonging to the John of Greenock, in
1818, happening to slip into a coil of running rope,
had his foot entirely cut off, and was obliged to have
the lower part of the leg amputated. A harpooner
belonging to the Hamilton, when engaged in lancing
a Whale, incautiously cast a little line under his foot.
The pain of the lance induced the Whale to dart suddenly downwards; his line began to run out from
under his feet, and in an instant caught him by a
turn round the body. He had just time to call out,
'Clear away the line.—Oh dear!' when he was
almost *cut asunder*, dragged overboard, and never
seen afterwards." Many such-like anecdotes are on
record.

When a boat is "fast" to the Whale, a little flag
is instantly hoisted in the stern as a signal to the
ship, and other boats are at once despatched to its
assistance. Sometimes, before their help can arrive,
the united lines of the boats first sent are all run
out, in which case the men are obliged to cut the
line, and lose it with the Whale, or the boat would
be dragged under water. But generally some of the
free boats can approach sufficiently near the animal
on his return to the surface, to dart another harpoon
into his body; perhaps he again dives, but returns

much exhausted. The men now thrust into his body long and slender steel lances, and aiming at the vitals these wounds soon prove fatal: blood mixed with water is discharged from the blow-holes, and presently streams of blood alone are ejected, which frequently drench the boats and men, and colour the sea far around. Sometimes the last agony of the victim is marked by convulsive motions with the tail, attended with imminent danger; but at other times, it yields its life quietly, turning gently over on its side. The flags are now struck, three hearty cheers resound, and the unwieldy prey is towed in triumph to the ship.

So huge a mass, of course, is slowly moved through the water, but there are few operations that are more-joyously performed; it is like the harvest-home of the farmer. When arrived, it is secured alongside the ship, and somewhat stretched by tackles at the head and tail, and the process of *flensing* commences. The men having shoes armed with long iron spikes to maintain their footing, get down on the huge and slippery carcass, and with very long knives and sharp spades make parallel cuts through the blubber, from the head to the tail. A band of fat, however, is left around the neck, called the *kent*, to which hooks and ropes are attached for the purpose of shifting round the carcass. The long parallel strips are divided across into portions weighing about half a ton each, and being separated from the flesh beneath, are hoisted on board, chopped into pieces, and put into casks. When the whalebone is exposed, it is detached by spades, &c., made for the purpose, and hoisted on deck in a mass; it is then split into junks, containing eight or ten blades each. Sometimes the jaws are taken out, and being fixed in a perpendicular position on deck, with the extremities in vessels, a considerable quantity of oil gradually drains from them. The carcass is then cut away as valueless to man, though a valuable prize to bears, birds, and sharks. Sometimes the carcass sinks immediately. Mr. Scoresby mentions a case in which it had been cut adrift prematurely, one of the men being still upon it; it began to sink, but unfortunately a hook in his boot had a firm hold of the flesh; he convulsively grasped the side of the boat in which his comrades were, and the whole immense weight was suspended by his foot. The torture was extreme; it was expected every instant that his foot would be rent off, or that his body would be torn asunder; but presently, by the merciful interposition of God, one of his companions contrived to hook a grapnel into the carcass, and it was drawn sufficiently near the surface for him to be extricated.

The Whale to which the preceding notices refer, is by no means the largest of the tribe, as the Great Rorqual (*Balænoptera boops*) sometimes attains nearly double the length of the former. Two specimens have been measured of the length of one hundred and five feet, and Sir A. de Capell Brooke asserts, that it is occasionally seen of the enormous dimensions of a hundred and twenty feet The Rorqual inhabits the same seas as the "right" Whale, but is not usually seen in company with it; they seem rather to avoid each other. The thinness of its blubber, and the shortness of its whalebone, render it of far less value than the other species; besides which, its swiftness, strength, and determination, render it a hazardous enemy to encounter. Hence it is usually avoided by the whalers, though the adventurous inhabitants of the Arctic shores of Europe do not hesitate to attack it. It is worthy of our notice, however, on account of its affording an instance of what has been called, in an examination of the care of Almighty God over his inferior creatures, the *principle of compensation*. When any organ, or set of organs that answer purposes very important in the economy of an animal, are removed in a kindred species with similar habits, or are so modified as no longer to serve the same purpose, some new structure is bestowed upon it, to supply the lack of that which is removed. We have seen how the Whale feeds, by receiving into its mouth a large quantity of water, which is filtered through the whalebone. In order to this, the mouth is made very capacious by the bowing over of the upper jaws in the form of a high arch, the blades of whalebone filling up the bow. But in the Rorqual the two jaws are nearly straight, and the blades vary little in length, so that thus far the cavity of the mouth is inconsiderable. Here comes in the compensation: the lower part of the mouth (or, externally, the chin and throat), instead of being stretched tightly across the branches of the lower jaw, are wrinkled up into many longitudinal folds, which, when the water rushes into the mouth, expand and make a capacious pouch or bag. On shutting the

mouth and contracting the muscles of the throat, the
flesh is pursed up again into folds, and the water is
driven, as in the former case, through the whalebone,
which secures the food.

The Whales, gigantic as they are, yet having little
power of offence, find to their cost, in common
with nobler creatures, that harmlessness is often no
resource against violence. Several species of the
voracious Sharks make the Whale the object of
their peculiar attacks; the Arctic Shark (*Scymnus
borealis*) is said, with its serrated teeth, to scoop out
hemispherical pieces of flesh from the Whale's body
as big as a man's head, and to proceed without mercy
until its appetite is satiated. Another Shark, often
called the Thresher (*Carcharias vulpes*), which is
sometimes upwards of twelve feet long, is said to
use its muscular tail, that is nearly half its whole
length, to inflict terrible slaps on the Whale; though
one would be apt to imagine that if this whipping
were all, the huge creature would be more frightened than hurt. The Sword-fish (*Xiphias gladius*),
however, in the long and bony spear that projects
from its snout, seems to be furnished with a weapon
which may reasonably alarm even the leviathan of
the deep, especially as the will to use his sword, if
we may believe eye-witnesses, is in nowise deficient.
The late Captain Crow records an incident of this
kind with much circumstantiality : "One morning,"
he observes, "during a calm, when near the Hebrides, all hands were called up at 3 A.M. to witness
a battle between several of the fish, called Threshers,
or Fox Sharks, and some Sword-fish, on one side,

SPERM WHALE ATTACKED BY SWORD-FISH.

and an enormous Whale on the other. It was in the middle of summer, and the weather being clear, and the fish close to the vessel, we had a fine opportunity of witnessing the combat. As soon as the Whale's back appeared above the water, the Threshers, springing several yards into the air, descended with great violence upon the object of their rancour, and inflicted upon him the most severe slaps with their long tails, the sound of which resembled the reports of muskets fired at a distance. The Sword-fish, in their turn, attacked the distressed Whale, stabbing from below; and thus beset on all sides and wounded, when the poor creature appeared, the water around him was dyed with blood. In this manner they continued tormenting and wounding him for many hours, until we lost sight of him; and I have no doubt they, in the end, completed his destruction."* Some discredit has been thrown on this and similar accounts, on the ground that the fishes could have no object in persecuting the Whale; but the circumstance is not more extraordinary than the well-known custom which little birds have of surrounding and teasing, or "mobbing," as it is called, any large bird to which they are unaccustomed. It has been objected, that the Captain describes the proceedings of the Sword-fish from below, when, from the reflection of the surface, he could not possibly see them. But, on the contrary, the incident is said to have occurred "close to the vessel;" and any one who has been at sea knows that in a calm, by going aloft, you can see to a great depth in the

* Memoirs of Capt. H. Crow, p. 11.

water. The habit here attributed to the Sword-fish
is confirmed by the frequency with which ships are
struck with great violence, most museums possessing
fragments of the planking of ships in which the
"sword" of this finny tyrant is imbedded. It is
with reason supposed that the dark and bulky hull
is by the fish mistaken for the body of a Whale.
The only resource which this gigantic animal has
for getting rid of his troublesome foes, is said to
be by diving to unfathomable depths, where their
structure could not for an instant sustain the enormous pressure.

Another animal has been accused of joining in
these assaults, I suppose from having been confounded with the Sword-fish. It is the Narwhal,
or Sea Unicorn (*Monodon monoceros*), a very different creature; in fact, being a first-cousin of the
Whale himself. This interesting animal, the beauty
of the northern seas, must be acquitted of this
charge, being as inoffensive as his great relative.
It is a very singular creature, formed in many respects like the Whale, but much more graceful.
The colour is grey above, and pure white beneath,
the whole spotted or mottled with a blackish hue.
From the head projects a long straight horn of solid
ivory, in the same line as the body; sometimes, but
rarely, there are two. The structure and origin of
this horn (which has given much celebrity to this
handsome creature) are very peculiar. It is, in fact,
the tooth, and the only one it possesses in general;
the fellow-tooth, however, exists within the bone of
the jaw, but undeveloped, lying shut up like the

SPEARING THE NARWHAL.

kernel of a nut. It is usually the left tooth that projects. Considerable uncertainty exists about the use of this long and spiral tusk. Some have supposed that it is used to search for food, by raking in the mud at the bottom, or to pierce thin ice at the surface, to obtain access to the air; but Mr. Scoresby appears to have thrown considerable light upon it, by having met with an individual in whose stomach, among the remains of other fishes, was found a skate, almost entire, which was two feet three inches long, and one foot eight inches wide. "Now it appears remarkable," observes this gentle-

man, "that the Narwhal, an animal without teeth,
a small mouth, and with stiff lips, should be able
to catch and swallow so large a fish as a skate,
the breadth of which is nearly three times as great
as the width of its own mouth. It seems probable
that the skates had been *pierced with the horn*, and
killed before they were devoured; otherwise it is
difficult to imagine how the Narwhal could have
swallowed them, or how a fish of any activity would
have permitted itself to be taken, and sucked down
the throat of a smooth-mouthed animal, without
teeth to detain and compress it."

We know but little of the true fishes that inhabit
the Arctic Seas. It appears, however, that many of the
more important of those which are common with us,
are common also there; not the subjects of an annual
migration, but widely distributed at all times. On
the authority of a French naval officer, it would even
seem that some species at least may undergo a sort
of torpidity. "Admiral Pleville Lepley, who had
had his home on the ocean for half a century, assured M. Lacépède that in Greenland, in the smaller
bays surrounded with rock, so common on this coast,
where the water is always calm, and the bottom
generally soft mud and juice, he had seen in the
beginning of spring myriads of Mackerel, with their
heads sunk some inches in the mud, their tails elevated vertically above its level; and that the mass
of fish was such, that at a distance it might be taken
for a reef of rocks. The Admiral supposed that the
Mackerel had passed the winter torpid, under the ice
and snow, and added that, for fifteen or twenty days

after their arrival, these fishes were affected with a kind of blindness, and that then many were taken with the net; but as they recovered their sight the nets would not answer, and hooks and lines were used."* In illustration of the great depth to which the eye can penetrate in these seas, from the transparency of the water, Captain Wood, who visited Spitzbergen in 1676, observed that, at the depth of four hundred and eighty feet, the shells on the bottom were distinctly visible.

The minute animals which constitute the food of the Whales, form a very interesting subject of contemplation. If any of my young readers have ever been upon the sea, though only in a boat, a few miles from the shore, they cannot fail to have observed floating in the water some round masses of transparent substance, like clear jelly, which alternately contract and dilate their bodies, or sometimes turn themselves, as it were, partly inside out. They are of various sizes, from that of a large plate to a microscopical minuteness; and some are set with rings, within each other, like eyes, and some are curiously fringed at the edge. These *Medusæ*, or Sea-blubbers, as they are familiarly called, form a considerable portion of the Whale's food, many species of them being abundant in its haunts. Another little animal occurs there in immense hosts, the *Clio borealis*, which bears some slight resemblance to a butterfly just emerged from the chrysalis, before the wings are expanded. Near the head there is on each side a large fin or wing, by the motions of

* Edin. Journal of Science.

which it changes its place. These ¬otions are
amusing; and as the little creatures are so abundant,

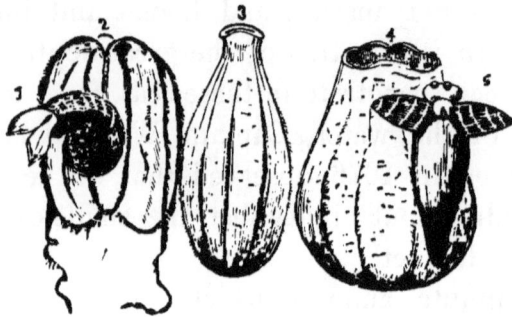

FOOD OF THE WHALE:

1, *Limacina helicina;* 2, 3, 4, *Meduæ;* 5, *Clio borealis.*

they make the dreary sea quite alive with their gambols as they dance merrily along.   In swimming, the
Clio brings the tips of its fins almost into contact,
first on one side, then on the other: in calm weather
they rise to the surface in myriads, for the purpose of
breathing but scarcely have they reached it before
they again descend into the deep. Mr. Scoresby
kept several of them alive in a glass of sea-water for
about a month, when they gradually wasted away
and died. The head of one of these little creatures
exhibits a most astonishing display of the wisdom of
God in creation. Around the mouth are placed six
tentacles, each of which is covered with about three
thousand red specks, which are seen by the microscope to be transparent cylinders, each containing
about twenty little suckers, capable of being thrust
out, and adapted for seizing and holding their minute
prey. "Thus, therefore, there will be three hundred

and sixty thousand of these microscopic suckers upon the head of one Clio; an apparatus for pre-,hension perhaps unequalled in the creation."

Numerous as are the hosts of these frolicsome little beings, there are, however, others which vastly exceed them in number; which pass, indeed, beyond the possibility of human computation. Navigators had often noticed, in certain parts of the Arctic Sea, that the water, instead of retaining its usual transparency, was densely opaque, and that its hue was grass-green, or sometimes olive-green. It is commonly known as the "green-water," and though liable to slight shiftings from the force of currents, is pretty constant in its position, occupying about one-fourth of the whole Greenland sea. Mr. Scoresby was the first who ascertained the cause of this peculiar hue: on examination he found that the water was densely filled with very minute *Medusæ*, for the most part undistinguishable without a microscope. He computes that within the compass of two square miles, supposing these animalcules to extend to the depth of two hundred and fifty fathoms, there would be congregated a number which eighty thousand persons, counting incessantly from the Creation until now, would not have enumerated, though they worked at the rate of a million per week! And when we consider that the area occupied by this green water in the Greenland seas is not less than twenty thousand square miles, what a vast idea does it give us of the profusion of animal life, and of the beneficence of Him who "openeth His hand, and satisfieth the desire of every living thing!"

Several species of minute Crabs and Shrimps occur also in great numbers, and constitute no small portion of the food of the Whale. One little creature, in particular (*Cancer nugax*), was found to swarm even beneath the ice, in the temporary sojourn of the discovery expeditions in winter quarters. The men had often noticed the shrinking of their salt meat which had been put to soak; and a goose that had been frozen, on being immersed to thaw, was, in the lapse of forty-eight hours, reduced to a perfect skeleton. The officers afterwards availed themselves of the services of these industrious little anatomists, to obtain cleaned skeletons of such small animals as they procured, merely taking the precaution of tying the specimen in a loose bag of gauze or netting, for the preservation of any of the smaller bones that might be separated by the consumption of the ligaments.

THE ATLANTIC OCEAN.

The Atlantic is much better known to us than any other of the great divisions of the Ocean, because, washing the shores of the principal commerical nations, it has been more traversed and explored. Its edges, on each side, are, in a greater degree than those of any other, hollowed into bays and harbours, and it is connected with the chief inland seas, such as the Baltic, Mediterranean, and Black Seas, on the one hand, and the Gulf of Mexico, and the Bays, or, rather Seas, of Hudson and Baffin, on the other. If, then, the importance of an Ocean is estimated by the length of the line of coast which borders it, the Atlantic takes precedency of all, exceeding even the Pacific in this respect, in the proportion of about four to three. It is remarkable, that it is the northern half which has so winding a coast, and to which, also, are confined the inland seas: and it is this part that is bordered with nations celebrated for navigation and commerce, the maritime nations of Europe and the United States. Unlike the Pacific, whose vast solitudes are rarely broken by the presence of a ship, the Atlantic is continually ploughed by the keels, and spangled with the banners, of powerful empires, conveying from shore to shore those diversified commodities, the interchange of which so

P   (169)

greatly promotes peace and good-will, and is, therefore, fraught with blessings to mankind.

Leaving behind us the inhospitable waters of the
north, let us take an imaginary voyage through this
important and interesting portion of the great deep,
still having an open eye to mark the footsteps of
Him whose "way is in the sea, and His path in the
great waters." The north breeze blows cheerily,
though coldly, and the sun, daily attaining a more
elevated position at noon, while the pole-star nightly
approaches the horizon, tells us of our rapid progress
southward. By and by, the shout of "Land ho!"
directs our attention to the horizon, where, with
straining eyes, we dimly discern what appears to be
a faint mass of cloud, of so evanescent a hue, that
a landsman looks long in the direction of the seaman's finger, and yet continues dubious whether
anything is really visible or not. Now he says con-
fidently, "Ha! I caught a glance of it then:" but
presently it turns out that his eye has been directed
to a point quite wide of the indicated locality; and
again he slowly but vainly sweeps the horizon with
his eye, in search of what the practised vision of the
mariner detects and recognises at a glance. Meanwhile, the ship rushes on before the cheerful breeze;
we go down to breakfast; and on again coming on
deck, there no longer remains any doubt; there lies
the land on the lee bow, high and blue, and palpable. It is one of the Azores; and as we draw
nearer, we discern and admire the picturesque beauties by which they are distinguished. The lofty
cliffs of varying hues rear their bold heads perpen-

dicularly from the foaming waves, cut and seamed into dark chasms and ravines, through which rocky torrents find a noisy course, while here and there a little stream is poured over the very summit of the precipice, the cascade descending in a white narrow line, conspicuous against the dark rock behind, until the wind carries it away in feathery spray, long before it reaches the bottom. The sunlight throws the prominences and cavities of the cliffs into broad masses of light and shadow, which, ever changing as the ship rapidly alters her position, give a magic character to the scene. Here and there, on the sides of the hills farther inland, the lawns and fields of lively green, speckled with white villas and hamlets, and relieved by the rich verdure of the orange-groves, present a softer but not less pleasing prospect. Other islands of this interesting group gradually rise from the horizon, all of similar character, but diverse in appearance from their various distance; some showing out in palpable distinctness, and others seen only in shadowy outline. But there is one which, from the singularity of its shape, arrests the attention. A mountain, of a very regularly conical form, seems to rise abruptly from the sea, with remarkable steepness, verdant almost to the summit; it is almost like a sugar-loaf, with a rounded top, crowned by a nipple-like prominence, which is often veiled by clouds. It is the Peak of Pico, seven thousand feet in height, second in celebrity, as in elevation, only to the Peak of Teneriffe. A recent visitor has thus described the picturesque beauty of this oceanic mountain:—" The hoary head

Pico.

of Pico presents a great variety of beauty. One
afternoon it was lightly powdered with snow, so as
to give it a tint of sober olive; with a larger quantity of frost or snow, and stronger and more direct
sunshine, it has looked like dead silver; at another
time it was tipped with fire; at another it was pavi-
lioned in flame-coloured clouds;—a few light mists
would shut it entirely out, or, where transparent,
give to it a wan and visionary hue; and in the evening, when the clouds put on a gayer livery, becoming
rose-coloured, or purple, or bronzed, the changes and
flushes would almost remind you of the variable
colours on a pigeon's neck; or, as a poet has said,

'Of hues that blush and glow
Like angels' wings.' "*

* Bullar's Azores, i. 368.

Some curious traditions are found in the writings of the ancients respecting an island of very large size, believed to have once existed in the Atlantic. Plato, in the Timæus, gives the fullest account of this island, which was called Atlantis. It is stated to have been nearly two hundred miles in length, situated opposite the Straits of Gibraltar. It was fertile and populous, and some of the warlike chiefs among whom it was divided, are said to have made irruptions upon the continent, and to have conquered a considerable part of Europe and Northern Africa. Several other islands are described as situated in the vicinity of Atlantis, beyond which lay a continent superior in size to all Europe and Africa. At length, the whole island is reported to have been swallowed up by the sea; after which, for a long period, that part of the Ocean was of difficult and dangerous navigation, on account of the numerous rocks and shelves which lay beneath the surface. There are many circumstances which render it improbable that this story, marvellous as it is, is entirely a fiction. It has been supposed that the great island was Cuba, the surrounding ones the other West Indies, and the great continent America; and that the cessation of intercourse with these regions, through the decay of naval enterprise, gave rise to the tradition that the island itself had disappeared. But this would not explain the matter-of-fact statement of the rocky shallows after the catastrophe; nor would the distance of Cuba from Europe permit martial invasions of this continent to be readily made from it. Others have concluded—and this does not seem to my own mind inconsistent with probability—that the statements of the ancients may be literally true; that by the action of an earthquake, of which we have had instances in modern times, the island may have been submerged, and that the Azores are the summits of the highest mountains. It seems somewhat to confirm this opinion, that these islands are evidently volcanic in their origin, and are very subject to earthquakes,—nay, the very phenomenon of islands swallowed up by the sea has repeatedly occurred here within historical record. It is true, that in these instances the island itself was small, and had been but recently raised by volcanic action; but it does not seem necessary that in similar cases there should be an exact parallelism, either in size or duration. The last of these occurrences was so remarkable on other accounts as to be well worthy of a detailed description, which is given by an eye-witness, Captain Tillard, an officer of the British navy : "Approaching the island of St. Michael's, on the 12th June, 1811, we occasionally observed, rising in the horizon, two or three columns of smoke, such as would have been occasioned by an action between two ships, to which cause we universally attributed its origin. This opinion was, however, in a very short time changed, from the smoke increasing and ascending in much larger bodies than could possibly have been produced by such an event; and having heard an account, prior to our sailing from Lisbon, that in the preceding January or February a volcano had burst out within the sea near St. Michael's, we immediately concluded that the smoke we saw proceeded from that cause, and on our anchoring the next morning in the road of Ponta del Gada, we found this conjecture correct as to the cause, but not as to the time; the eruption of January having totally subsided, and the present one having only burst forth two days prior to our approach, and about three miles distant from the one before alluded to."

The Captain having proceeded to a cliff on the island of St. Michael's, about three or four hundred feet high, from which the eruption was scarcely a mile distant, proceeds to describe its appearance: "Imagine an immense body of smoke rising from the sea, the surface of which was marked by the silvery rippling of the waves. In a quiescent state, it had the appearance of a circular cloud revolving on the water, like a horizontal wheel, in various and irregular involutions, expanding itself gradually on the lee side; when suddenly, a column of the blackest cinders, ashes, and stones, would shoot up in the form of a spire, at an angle of from ten to twenty degrees from a perpendicular line, the angle of inclination being universally to windward; this was rapidly succeeded by a second, third, and fourth shower, each acquiring greater velocity, and overtopping the other, till they had attained an altitude as much above the level of our eye as the sea was below it.

"As the impetus with which the several columns were severally propelled diminished, and their ascending motion had nearly ceased, they broke into various branches resembling a group of pines: these

SUBMARINE VOLCANO.

again forming themselves into festoons of white feathery smoke, in the most fanciful manner imaginable,
intermixed with the finest particles of falling ashes,
which at one time assumed the appearance of innumerable plumes of black and white ostrich feathers
surmounting each other; at another, that of the
light wavy branches of a weeping willow.

"During these bursts, the most vivid flashes of
lightning continually issued from the densest part of
the volcano; and the cloud of smoke now ascending to an altitude much above the highest point to
which the ashes were projected, rolled off in large
masses of fleecy clouds, gradually expanding themselves before the wind in a direction nearly hori-

zontal, and drawing up to them a quantity of water-spouts, which formed a most beautiful and striking addition to the general appearance of the scene."

In the course of a few hours, a crater had been thrown up by these eruptions, to the height of twenty feet above the sea, and apparently three or four hundred feet in diameter. Repeated shocks of an earthquake accompanied the explosion. The narrator was obliged to leave the neighbourhood on the succeeding day, at which time the volcanic eruption was seen from a distance to be still raging with undiminished fury. About three weeks afterwards he returned to the spot, and found all quiet, but the newly-formed island had increased to a mile in circumference, and the highest part appeared to have an elevation of about two hundred and forty feet. On landing, he found the place still smoking, and the larger crater nearly full of water in a boiling state, which was being discharged into the Ocean by a stream about six yards across: this stream, close to the edge of the sea, was so hot, as barely to admit the momentary immersion of the finger.* On the 11th of October, in the same year, this island sank beneath the Ocean from which it had emerged, leaving a dangerous shoal in the neighbourhood, thus realizing the traditionary fate of the island of Atlantis.

But let us pursue our voyage. As we follow the setting sun to his bed among the Indian islands of the west, the tedium of our way across the trackless

* Trans. Roy. Soc. 1812.

waste is enlivened by those cheerful little birds, the
Petrels (*Procellaria pelagica*), the constant companions of the sailor, by whom they are familiarly
named Mother Carey's chickens. They are peculiarly Ocean-birds: rarely approaching the shore,
except when they seek gloomy and inaccessible rocks
for the purpose of breeding; they are never seen but
in association with the boundless waste of waters.
Scarcely larger than the swallow that darts through
our streets, one wonders that so frail a little bird
should brave the fury of the tempest; but when the
masts are cracking, and the cordage shrieking fit-
fully in the fierce blast, and when the sea is leaping
up into mountainous waves, whose foaming crests
are torn off in invisible mist before the violence of
the gale, the little Petrel flits hither and thither,
now treading the brow of the watery hill, now
sweeping through the valley, piping its singular note
with as much glee as if it were the very spirit of the
storm, which the superstitious mariner, indeed, attributes to its evil agency. Flocks of these little birds,
more or less numerous, accompany ships, often for
many days successively, not, as has been asserted,
to seek a refuge from the storm in their shelter,
but to feed on the greasy particles which the cook
now and then throws overboard, or the floating substances which the vessel's motion brings to the surface. It is a pleasing sight to see them crowd up
close under the stern with confiding fearlessness,
their sooty wings horizontally extended, and their
tiny web-feet put down to feel the water, while they
pick up with their beaks the minute atoms of food

of which they are in search. I have been surprised
to notice how very quickly a flock will collect,
though a few moments before scarcely one could
be seen in any direction; and again they disperse
as speedily. They seem to have the power of dis-
pensing with sleep, at least for very long intervals.
Wilson, one of the most accurate of observers, has
recorded a fact illustrative of this: "In firing at
these birds, a quill-feather was broken in each wing
of an individual, and hung fluttering in the wind,
which rendered it so conspicuous among the rest, as
to be known to all on board. This bird, notwithstanding its inconvenience, continued with us for
nearly a week, during which we sailed a distance
of more than four hundred miles to the north." Of
course, if this individual had gone to sleep, the
vessel would have sailed away, and we can hardly
imagine that it would have again found her in her
pathless course. I do not believe they have ever
been known to alight on the rigging or deck of a
ship.

It is a pity that so interesting a little creature
as this should become the object of a degrading
and meaningless superstition. The persuasion that
they are in some mysterious manner connected with
the creation of storms, is so prevalent among seamen, as to render them, innocent and confiding as
they are, objects of general dislike, and often even
of hatred. I once made a voyage with a captain,
who, though a man of much intelligence, was not
proof against this absurd superstition, venting hearty
execrations against these "devil's imps," as he called

them, in every gale, as if they had been the mali-
cious authors of it.   If this unoffending little bird
does afford any indication of a coming storm, discovered by its more acute perceptions, which, never-
theless, I very much doubt, why should not those
who navigate the Ocean, receive its warning with
gratitude, and make preparations for security, instead
of following it with profane and impotent curses?
" As well might they curse the midnight lighthouse
that, star-like, guides them on their watery way, or
the buoy that warns them of the sunken rocks below,
as this harmless wanderer, whose manner informs
them of the approach of the storm, and thereby
enables them to prepare for it."

   A frequent relief to the tedium of a long voyage
is found in the shoals of playful Dolphins (*Del-
phinus delphis*, &c.) which so often perform their
amusing gambols around us.   They may be discerned
at a great distance; as they are continually leaping
from the surface of the sea, an action which, as it
seems to have no obvious object, is probably the
mere exuberance of animal mirth.   When a shoal is
seen thus frolicing at the distance of a mile or two,
in a few moments, having caught sight of the ship,
down they come trooping with the velocity of the
wind, impelled by curiosity to discover what being
of monstrous bulk thus invades their domain.   When
arrived, they display their agility in a thousand
graceful motions, now leaping with curved bodies
many feet into the air, then darting through a wave
with incredible velocity, leaving a slender wake of
whitening foam under the water; now the thin back-

fin only is exposed, cutting the surface like a knife;
then the broad and muscular tail is elevated as the
animal plunges perpendicularly down into the depth,
or dives beneath the keel to explore the opposite
side. So smooth are their bodies, that their gambols are performed with surprisingly little disturbance
of the water, and even when descending from their
agile somersets they make scarcely any splash. The
colour of the upper parts of their bodies is of a deep
black, but by a deception of the sight, caused, probably, by the swiftness of their motions, and by the
gleaming of the light from their wet and glittering
skin, they appear in the air and under water of a
light-greenish grey. After having taken a few rapid
turns under and around the vessel, the whole shoal,
consisting of a dozen or two, usually congregate
immediately beneath the bowsprit, where they remain sometimes for hours, romping and rolling about
as if the ship were perfectly stationary, instead of
spanking along at the rate of seven or eight knots
an hour, apparently making no effort to go ahead,
and yet keeping their relative position with admirable dexterity and precision. But they are allowed
to remain so long undisturbed only when the duties
of the ship demand the attention of the hands: for
if there be a few moments of leisure, the presence
of a shoal of Dolphins is too tempting to pass un-
heeded. Some one of the crew reputed to be skilful in wielding the harpoon, in small vessels often
the captain himself, goes forward, and having taken
his station upon the bowsprit-heel, or upon one of
the cat-heads, poises his implement of war, and waits

Q

a favourable moment of attack. Now the bows are
thronged with anxious faces; the usual discipline
of the ship is relaxed on such occasions; even the
sooty cook leaves his caboose, and with the dirty
cabin-boy endeavours to witness the interesting per-
formance. All are there but the man at the wheel,
and even he stands on tip-toe to catch a glimpse
of what is going on, and neglecting his helm, "yaws"
the ship about sadly. The unsuspecting visitors
continue their romps: presently one comes within
aim, pretty near the surface; the dart is thrown, and
if the trembling anxiety of the harpooner have not
marred his skill, strikes its object: I have known
it, however, take effect obliquely on the side, cutting
deeply into the flesh, but retaining no hold; in which
case the poor wounded creature, with its bowels exposed and protruding, instantly shoots away, accompanied by all its fellows, not, however, to sympathize
with it, or afford it any assistance, but, if the sailors
may be believed, to fall upon and devour it. But
we will suppose that the barbed weapon has transfixed the animal in the back, and, piercing through
the superficial coat of fat, has lodged deep in the
solid flesh. The Dolphin plunges convulsively: the
whole herd are gone like a thought, leaving their
unhappy comrade to his fate: the stout line stretches
with the force, but brings him up with a jerk; the
barbs are beneath the tough muscles, and resist all
his endeavours for freedom: a dozen eager hands
are thrust forth to grasp the line and haul him to
the surface. The struggles of the desperate creature are now tremendous: the water all around is

lashed into boiling foam, reddened with the life-blood that is fast ebbing from his wound. Two or three of the most agile now jump into the fore-chains, with the end of a rope formed into a running noose; they hang this down into the water, and endeavour to get the bight over his tail; many trials are un-successfully made to do this, for the frantic motions of the animal render it a very difficult operation; at length, however, it is drawn over, tightened, and the prey is considered secure. It is now comparatively easy, with the aid of a boat-hook, to pass another rope under the body, just behind the breast-fins, and then he is soon hoisted on deck. I have been asto-nished to observe how very inadequate is the notion one forms of the dimensions of these animals by see-ing them only in the water; an individual that mea-sures eight feet in length, appearing in water not more than four or five. The muscular power is very great, but is chiefly concentrated in the tail, and, therefore, when the animal is removed from its na-tive element, it is almost helpless, its exertions being confined to the violent blows which it inflicts upon the deck with this broad and powerful organ. In all essential particulars, the Dolphin agrees with the Whale already described, being of the same order; but it differs in having an upright fin on the back, and both the upper and lower jaws armed with nume-rous small, close, and pointed teeth. In one speci-men which I saw captured, I counted one hundred and fifty-two in all; they are beautifully regular, and those of one jaw fit into the interstices of the other. The Dolphin differs from the Porpesse (*Pho-*

cæna) by having the jaws lengthened out into a long
and slender beak, almost like that of some bird: in
other respects, there is little difference between the
Porpesse and the Dolphin. Both are very voracious,
pursuing any prey they can master: in the stomach
of one taken in the Atlantic, I found a number of
the beaks of Cuttles (*Sepiadæ*). A century or two
ago, the flesh of this animal was esteemed a dainty
worthy the attention of epicures in this country;
but now it is relished only by those whom the salt
provisions of a long voyage have rendered less choice
than they would be under other circumstances. From
the abundance of blood, the meat is very dark in
appearance; but to my own taste, on one or two
occasions, with my appetite sharpened by the privation just mentioned, steaks cut from it and fried
have seemed very savoury and agreeable.

  Now the long yellow strings of floating weed,
which lie in parallel lines pointing to the wind, or
the broader masses that resemble meadows parched
by protracted drought, inform us that we are in that
mighty current of tepid water, the Gulf-stream. We
hasten to the gangway, and having drawn a few
buckets of clear transparent water, which we deposit
in a tub, collect with a boat-hook, a quantity of the
floating weed, and immerse it in the tub of water
to be examined. Many of the stems and berry-
like air-vessels are coated with a thin and delicate
tissue of shelly substance (*Flustra*), of a greyish
hue, like very minute network, so delicate as not
at all to disfigure or conceal the form of the substance on which it is spread. Attached to the weed

are groups of little Barnacles (*Lepas*), from the size of a pin's head to half an inch in length. While under water, these are incessantly projecting and retracting the elegant curled apparatus of *cirri* with which they are furnished, resembling a plume of feathers; from which resemblance it probably was that the inhabitants of a species found on the Scottish coast were asserted to be "of that nature to be finally by nature of seas resolved into geese."* The purpose of this continual motion of the fringed arms appears to be twofold; first, to make a constant eddy in the surrounding water, and thus bring minute animals within reach, and then to enclose such as are brought in as by the cast of a net, and convey them to the mouth. Crawling on the surface of the weed we may now and then find a nimble little Crab (*Lupa*), with the shell on each side projecting horizontally into a sharp spine. We are surprised at first to find a Crab on the surface of the Ocean, as the species with which we are familiar have not the power of swimming. On endeavouring to procure one for examination, however, we no sooner touch the fragment of the weed with the boat-hook, than the watchful little Crab hurries off into the water, and swims rapidly away out of reach. If we be fortunate enough to secure one by skilful manœuvring with the bucket or dip-net, we shall discover a peculiar structure, by means of which these Ocean-crabs are endowed with the faculty of swimming. In the common Crab, all the feet, except the claws, terminate in a sharp point, but in the present genus

* Boëce, Cosmography of Albioun. Edin. about 1541.

the hindmost pair have the last joint flattened out into a thin but broad oval plate, the edge of which is thickly fringed with fine hairs. This structure is exactly parallel to that by which the foot of a perching bird is modified into the foot of a swimming bird, the surface being dilated into a broad web; or to the wide fringe by which the hind feet of a water-beetle are made such powerful oars; the flattened joint in the present case becoming a paddle, by the stroke of which a rapid motion is obtained through the water. These Swimming Crabs are very voracious, preying upon the little shrimps that are numerous about the weed, which they pursue and seize with their pincers. Sometimes the Crab remains at rest, but vigilant, until a shrimp swims within reach, when he grasps it with great quickness, and proceeds to devour it by degrees. In doing this, he holds it fast by one claw, while with the other he picks off very daintily the legs and other members of his prey, putting them bit by bit into his mouth, until nothing remains but the tail, which he rejects.

The weed is usually the resort of several small species of fishes, which doubtless congregate about it for the sake of the minute Crustacea that are so abundant. Among them I have found a very interesting little species of Toad-fish (*Antennarius*), whose pectoral and ventral fins project so far from the surface of the body as to expose the joint, and thus take the form of the feet of a quadruped. It uses these members actually as feet, crawling and pushing its way among the tangled weed by means of them.   It has even been known to come on shore, and remain several days without any communication with the water.   On the head of this fish there are one or two slender horns, furnished at the tip with several processes resembling little worms.   The use of these organs is very remarkable.   The fish is not one of swift motion, and therefore cannot take its prey by pursuit: instead of this, it usually conceals itself among the mud at the bottom, or perhaps among the stalks of floating weed, while it agitates its curious fleshy horns; their resemblance to worms and their motion attract other fishes, which, coming within reach, are seized by the capacious mouth of the latent Toad-fish.   The lower jaw extending beyond the upper, causes the mouth to open perpendicularly, and the eyes are so situated as to look in the same direction, both of which arrangements facilitate the capture of prey by this singular mode.   It is not improbable that the worm-like tentacles attached to the mouth and chin of other fishes, as the Cod and Barble, for example, answer an end somewhat similar to this.

In keeping small marine animals for examination, we often lose the specimens through the water becoming speedily unfit for supporting animal life; a minute Shrimp or two, or a fish of an inch in length, if confined in a large basin of water, will usually exhaust the oxygen during the night, and be dead by the morning.   A little living seaweed, however, placed with them, will prevent, or, at least, delay this, as plants in a living state give out oxygen.

Every night the pole-star is perceptibly nearer the

horizon, and every day the meridian sun reaches to
a higher and yet a higher point, until it appears almost vertical. The wind gradually becomes lighter,
until we arrive at the "calm latitudes," where we
lie weeks without making any progress. The captain and crew *whistle for wind* with as much per-
severance as if they had never been disappointed,
and every one watches anxiously for the least breath-
ings of a breeze. Nothing can exceed the tantaliz-
ing tedium of this condition; the wearied eye gazes
intently upon the glistening sea, and eagerly catches
the slightest ruffling of the mirror-like smoothness,
in hopes that it may be an indication of wind; but
on glancing at the feather-vane upon the ship's quarter, the hope fades on perceiving it hang motionless
from its staff. A still more delicate test is then resorted to, that of throwing a live coal overboard,
and marking if the little cloud of white steam has
any lateral motion; but no! it ascends perpendicularly till dispersed in the air. Now and then,
the polished surface of the sea is suddenly changed
to a blue ripple; expectation becomes strong, for
there is no doubt of the reality of the motion; but
before the sails can feel the breeze, it has died away
again; the air is as still, and the sea as glassy, as
before. Coleridge has well described such a state in
his "Ancient Mariner:"—

> "The sun came up upon the left,
>   Out of the sea came he;
> And he shone bright and on the right
>   Went down into the sea.

*　　*　　*　　*　　*　　*

"Down dropp'd the breeze, the sails dropp'd down;
 'Twas sad as sad could be:
 And we did speak only to break
 The silence of the sea.

"Day after day, day after day,
 We stuck, nor breath nor motion;
 As idle as a painted ship
 Upon a painted ocean."

Not a cloud tempers the fierce burning rays of the sun, which shoot directly on our heads; the deck becomes scalding hot to the feet, the melting pitch boils up from the seams, the tar continually drops from the rigging, the masts and booms display gaping cracks, and the flukes of the anchors are too hot to be touched with impunity. In vain, if we happen to be sailing in a small vessel, which has no awning on board to spread over the quarter-deck, we seek for refuge beneath the sails which hang lazily from the yards and gaffs, inviting the desired gales; for so perpendicular are the fiery beams in the heat of the day, that very little shadow is afforded by the sails, and even that little is constantly shifting from the vessel's change of position in the swell. In such circumstances, I have in some measure felt the force of those similitudes in the Sacred Prophets, in which the blessings of the coming reign of the Lord Jesus Christ, after the long apostacy, are likened to "the shadow of a great rock in a weary land." "Thou hast been a shadow from the heat, when the blast of the terrible ones is as a storm against the wall. Thou shalt bring down the noise of strangers, as the heat in a dry place; even the heat with the shadow of a cloud."*

Yet, though day after day rolls on and leaves us still in the same position, there are not wanting many things to beguile the weariness of the time. The gorgeous beauty of the sun's setting almost makes amends for his unmitigated heat by day. As his orb approaches the western horizon, the clouds, which have been absent during the day, begin to form in that quarter of the heavens; and, as he sinks, assume hues of the richest purple edged with gold, now hiding his disc, now allowing him to flash out his softened effulgence through crimson openings, till he falls beneath the massy mountain-like bed of cloud that seems to lie heavily upon the surface of the sea. Then the whole array begins to take the appearance of a lovely landscape; the clouds forming the land, while the open sky represents calm water. Sometimes we seem to see the long capes and bold promontories of a broken and picturesque coast, deeply indented with bays and creeks, and fringed with groups of islands; at others, silvery lakes, studded with little wooded islets, appear embosomed in mountains or surrounded by gentle slopes, here and there clothed with umbrageous woods. Such an appearance of reality is given to these fleeting scenes, that it is difficult, after gazing at them for a few minutes, to believe they are mere shadows. The mind forgets the world of waters around, and, in the enthusiasm of the hour, goes out in busy imagination to that beautiful land, and roves among

* Isa. xxxii. 2; xxv. 4, 5; iv. 6.

its valleys and hills in dreamy enjoyment. We are not, then, surprised that the imaginative Greeks should have sung of their Fortunate Islands, the habitations of the blessed, placed far away in the ocean of the west, and invested with more than earthly loveliness; nor that the existence of isles of similar character, in the same mysterious, because unknown, regions, should have found a place in the mythology of even so remote a nation as the Hindoos.

The beauteous scenes before us, however, are as transitory as they are lovely: night comes on with a rapidity, startling to us accustomed to the long twilight of the north; the rich hues with which the western sky is suffused, the crimson and ruddy gold, speedily change to a warm and swarthy brown, and one by one the stars come out, and light up the sky with a strange and unwonted effulgence. Humboldt describes in the following terms his own emotions on first seeing the brilliant stars of these regions :—

"From the time we entered the torrid zone, we were never wearied with admiring, every night, the beauty of the southern sky, which, as we advanced towards the south, opened new constellations to our view. We feel an indescribable sensation, when, on approaching the equator, and particularly on passing from one hemisphere to the other, we see those stars which we have contemplated from our infancy, progressively sink, and finally disappear. Nothing awakens in the traveller a livelier remembrance of the immense distance by which he is separated from his country, than the aspect of an

unknown firmament. The grouping of the stars of
the first magnitude, some scattered nebulæ rivalling
in splendour the milky way, and tracts of space
remarkable for their extreme blackness, give a particular physiognomy to the southern sky. This
sight fills with admiration even those, who, unin-
structed in the branches of accurate science, feel
the same emotions of delight in the contemplation
of the heavenly vault, as in the view of a beautiful
landscape, or a majestic river. A traveller has no
need of being a botanist to recognize the torrid zone
on the mere aspect of its vegetation; and, without
having acquired any notions of astronomy, he feels
he is not in Europe, when he sees the immense con-
stellation of the Ship, or the phosphorescent clouds
of Magellan, arise on the horizon. The heaven and
the earth, everything in the equinoctial regions, assume an exotic character."*

But of all the constellations that stud the sky of
the southern hemisphere, there is none that more
strikes a stranger than the Southern Cross. Its
beauty, as well as the singularity of its form, cannot
fail to inspire interest; even though we be, through
the grace of God, furnished with ideas of true
and spiritual worship, that prevent our viewing it
with the superstitious reverence with which it is
regarded by the inhabitants of South America. It
is not seen above the horizon until we are within
the tropics, and scarcely appears to advantage until
we approach the equator. As the two brilliant stars
which form the top and bottom of the Cross, have

* Personal Narrative, 1814. Vol. ii. p. 18.

nearly the same right ascension, they assume a perpendicular position when upon the meridian; and hence afford an accurate mode of measuring time;

THE SOUTHERN CROSS.

as the hour of *southing* at the different seasons, varying four minutes every night, is well known to the inhabitants of the southern hemisphere. It is very common to hear the peasants observe one to another, "It is after midnight" (or some other hour); "the Cross begins to fall!"

Alone, in the midst of the ocean, called to nightly watchings upon the deck, the mariner naturally becomes familiar with the glowing orbs which are revealed by the surrounding darkness; and if he be a Christian, his thoughts are led out, as he lifts

13   R

up his eyes on high, and beholds the stars marshalled in order, or the moon "walking in brightness," to
Him that "created these things, that bringeth out
their host by number, and calleth them all by
names." For "the heavens declare the glory of
God; and the firmament sheweth His handywork.
Day unto day uttereth speech; and night unto night
sheweth knowledge. There is no speech nor language, where their voice is not heard."

Between, or in the neighbourhood of the tropics,
the ship is rarely unaccompanied by fishes of many
species, which, in the clear waters of these southern
seas, are visible many fathoms beneath her keel.

CORYPHENE (*Coryphœna*).

One of the most common, and perhaps one of the
most beautiful, is the Coryphene (*Coryphœna*), mis-
called by seamen, the Dolphin. One is never weary

of admiring their beauty. Their form is deep, but thin and somewhat flattened; and their sides are of brilliant pearly white, like polished silver. In small companies of five or six, they usually appear and play around and beneath the ship, sometimes close to the surface, and sometimes at such a depth that the eye can but dimly discern their shadowy outline. When playing at an inconsiderable depth, in their turnings hither and thither, the rays of the sun, reflected from ther polished sides, as one or the other is exposed to the light, flash out in sudden gleams, or are interrupted, in a very striking manner. Night and day these interesting creatures are sporting about, apparently insusceptible of weariness. Their motion is very rapid, when their powers are put forth, as in pursuit of the timid little Flyingfish. It is to these fishes that most of the accounts of Dolphins, which we read in voyages, must be referred, as, owing to some mistake of identity, not easily accounted for, the name of Dolphin has been universally misapplied by our seamen to the Coryphene, while they confound the true Dolphin with the Porpesse. From not adverting to this habitual misnomer, some confusion has arisen: thus the following interesting notice has been quoted in a late valuable work on the Cetacea,* as illustrative of the true Dolphins, although the fair narrator herself takes care to inform us that she means the *Coryphæna hippuris:* "The other morning, a large Dolphin, which had been following the ship for some distance, and was sparkling most gloriously

* Jardine's Naturalist's Library.

in the sun, suddenly detected a shoal of Flying-fish
rising from the sea at some distance. With the
rapidity of lightning he wheeled round, made one
tremendous leap, and so timed his fall as to arrive
fairly at the place where our little friends, the Flying-fish, were forced to drop into the sea to refresh
their weary wing. A flight of sea-gulls now joined
in the pursuit; we gave up our protégés for lost,
when, to our great joy, we beheld them rising again,
for they had merely skimmed the wave, and, thus
recruited, continued their flight. Their restless foe
pursued them with giant strides, now cutting the
wave, which flashed and sparkled with the reflection
of his brilliant coat, and then giving one huge leap,
which brought him up with his prey: they seemed
conscious that escape was impossible; their flight
became shorter and more flurried, whilst the Dolphin,
animated by the certain prospect of success, grew
more vigorous in his bounds; exhausted, they dropped their wings, and fell one by one into the jaws
of the Dolphin, or were snapped up by the vigilant
Gulls.''*

Captain Basil Hall has described a very similar
scene in nearly parallel terms; but, to prevent mis-
understanding, he also informs his readers that "the
Dolphin" of his narrative is the *Coryphæna hippuris*
of naturalists, and a true fish.

"Shortly after observing a cluster of Flying-fish
rise out of the water, we discovered two or three
Dolphins [Coryphenes] ranging past the ship, in all
their beauty; and watched with some anxiety to

* Miss Lloyd's Sketches of Bermuda.

PURSUIT OF FLYING-FISH BY DOLPHINS AND BIRDS.

see one of those aquatic chases, of which our friends
the Indiamen had been telling us such wonderful
stories. We had not long to wait; for the ship,
in her progress through the water, soon put up
another shoal of these little things, which, as the
others had done, took their flight directly to windward. A large Dolphin, which had been keeping
company with us abreast of the weather gangway,
at the depth of two or three fathoms, and, as usual,
glistening most beautifully in the sun, no sooner
detected our poor, dear little friends take wing, than
he turned his head towards them, and, darting to
the surface, leaped from the water with a velocity
little short, as it seemed, of a cannon-ball. But,
although the impetus with which he shot himself
into the air gave him an initial velocity greatly
exceeding that of the Flying-fish, the start which his
fated prey had got, enabled them to keep ahead of
him for a considerable time.

"The length of the Dolphin's first spring could
not be less than ten yards; and, after he fell, we
could see him gliding like lightning through the
water for a moment, when he again rose and shot
forwards with considerably greater velocity than at
first, and, of course, to a still greater distance. In
this manner the merciless pursuer seemed to stride
along the sea with fearful rapidity, while his brilliant coat sparkled and flashed in the sun quite splendidly. As he fell headlong on the water, at the end
of each huge leap, a series of circles were sent far
over the still surface, which lay as smooth as a
mirror.

"The group of wretched Flying-fish, thus hotly pursued, at length dropped into the sea; but we were rejoiced to observe that they merely touched the top of the swell, and scarcely sunk in it; at least, they instantly set off again in a fresh and even more vigorous flight. It was particularly interesting to observe, that the direction they now took was quite different from the one in which they had set out, implying but too obviously that they had detected their fierce enemy, who was following them with giant steps along the waves, and now gaining rapidly upon them. His terrific pace, indeed, was two or three times as swift as theirs, poor little things!

"The greedy Dolphin, however, was fully as quick-sighted as the Flying-fish which were trying to elude him; for, whenever they varied their flight in the smallest degree, he lost not the tenth part of a second in shaping a new course, so as to cut off the chase; while they, in a manner really not unlike that of the hare, doubled more than once upon their pursuer. But it was soon too plainly to be seen that the strength and confidence of the Flying-fish were fast ebbing. Their flights became shorter and shorter, and their course more fluttering and uncertain, while the enormous leaps of the Dolphin appeared to grow only more vigorous at each bound. Eventually, indeed, we could see, or fancied that we could see, that this skilful sea-sportsman arranged all his springs with such an assurance of success, that he contrived to fall, at the end of each, just under the very spot on which the exhausted Flying-fish were about to drop! Sometimes this catastrophe took place at too great a distance for us to see from the deck exactly what happened; but on our mounting high into the rigging, we may be said to have been in at the death; for then we could discover that the unfortunate little creatures, one after another, either popped right into the Dolphin's jaws as they lighted on the water, or were snapped up instantly afterwards.

"It was impossible not to take an active part with our pretty little friends of the weaker side, and accordingly we very speedily had our revenge. The middies and the sailors, delighted with the chance, rigged out a dozen or twenty lines from the jib-boom end and spritsail-yard-arms with hooks, baited merely with bits of tin, the glitter of which resembles so much that of the body and wings of the Flying-fish, that many a proud Dolphin, making sure of a delicious morsel, leaped in rapture at the deceitful prize."*

Though these and other recorded anecdotes indubitably refer to the bright pearly fishes just described, there cannot be a doubt that the same habits are found to mark the true Cetaceous Dolphins; while at the same time I confess that I do not recollect any instance in which such pursuit has been witnessed, in my own experience, or recorded in books of voyages. Indeed I do not conceive that the chase of the Flying-fish by the Coryphene has been *often* witnessed, nor that it can be considered as any other than a rare occurrence. As the aerial boundings of the Flying-

* Frag. Voy. and Trav. Second Series. Vol. i. p. 224.

fish, however, are of constant observation within the
tropics, it seems but natural to conclude that they
are but the frolicsome putting forth of superabundant
animal energy; that they are, in fact, performed in
sportive play, as the lamb skips and leaps upon the
grass, or the dog pursues its own evasive tail.
These flights, generally performed in shoals varying
in number from a dozen to a hundred or more, are
extremely pleasing, and sustain our interest even
long after they have become familiar to us. One
is apt, at first sight of a flock, especially if it be
unexpected, to mistake them for white birds flying
by, till they are seen to alight in the water. The
length of the bound is enormous, if it be indeed
effected by a single impulse; but this point seems
hardly to be satisfactorily settled even yet. I feel
persuaded that I have more than once seen them
deviate from the uniform curve which they usually
describe, rising and sinking alternately so as to
keep at the same distance from the undulations of
the surface; and Humboldt, one of the most accurate of observers, speaks unhesitatingly of their flap-
ping the air with their long fins. Indeed, it would
else seem almost impossible to imagine that so small
a fish, not so large as a herring, should be able to
propel itself to the height of twenty feet, and to the
distance of more than six hundred, through the air.
Generally, one takes his leap first, then the whole
flock follow at once, shooting in nearly a straight
line, and skimming along a little above the surface;
so little that they often strike the side of a rising
wave, and go under water.

Another visitant, who very freely gives us much
of his company, is the White Shark (*Carcarius vulgaris*), probably the most terrific monster that cleaves
the waves; certainly the most hated, and at the same
time feared, by the sailor. The catching of fish is at
all times a pleasing amusement to the mariner; but to
catch the "Shirk," as he is called, there is a peculiar
avidity, in which the gratification of a deep-seated
hatred of the species, and vengeance for his murder-
ous propensities, form the leading features. When
taken, whether entrapped by the concealed hook, or
struck by the open violence of the harpoon, and
brought on deck, he is subjected to every indignity
which an insane fury can heap upon an object—beat,
stabbed, and kicked, and even reviled as if capable
of understanding language. In truth, I have never
seen any animal, terrestrial or aquatic, which, so to
speak, has "villain" written on its countenance in as
legible characters as the Shark. The shape of the
head, and the form of the mouth, opening so far beneath, are anything but prepossessing; but there is
a peculiar malignity in the expression of the eye, that
seems almost satanic, and which one can never look
upon without shuddering. The mouth is armed with
teeth of very peculiar construction; they are triangular in form, thin and flat, the central part, however,
being thicker than the edges, which are as keen as a
lancet, and cut into fine serratures, like a saw. In
very large Sharks, the teeth have been found nearly
two inches in breadth: they are placed in rows,
sometimes to the number of six, one within another,
lying nearly flat when not in use, but erected in a

THE OCEAN.

moment to seize prey: and as they are so planted in
the jaw that each tooth is capable of independent
motion, being furnished with its own muscles, and as
the power of the jaws is enormous, they form one of
the most terrific and formidable apparatus existing
for the supply of carnivorous appetite. The fatal
voracity of this animal is well known: instances are
numerous of swimmers in tropical seas having been
severed in twain at one snap, or deprived of limbs,
while, on more than one occasion, the whole body of
a man has been taken from this living sepulchre.
Yet this sanguinary voracity is but the result of an
unerring instinct implanted in the animal by GOD,
without the exercise of which its life could not be
sustained: and therefore it seems not only foolish, but
even sinful, to entertain feelings of personal revenge
against it, as if it were endowed with human reason,
" knowing good and evil." I do not know that it is
wrong to kill an animal so destructive and dangerous; I reprobate only the imputation to it of
human motives, and the staining a useful act with
unnecessary cruelty.

The mode by which the race of these formidable
creatures is continued, differing as it does so greatly
from that of most other fishes, is exceedingly curious.
The Shark, instead of depositing some millions of
eggs in a season, like the Cod or the Herring, produces two eggs, of a square or oblong form, the coat
of which is composed of a tough horny substance;
each corner is prolonged into a tendril, of which the
two which are next the tail of the enclosed fish are
stronger and more prehensile than the other pair.

The use of these tendrils appears to be their entanglement among the stalks of sea-weeds, and the consequent mooring of the egg in a situation of protection and comparative security. Near the head there is a slit in the egg-skin, through which the water enters for respiration, and another at the opposite extremity by which it is discharged. That part of the skin which is near the head, is weaker and more easily ruptured than any other part; a provision for the easy exclusion of the animal, which takes place before the entire absorption of the *vitellus* or yolk of the egg, the remainder being attached to the body of the young fish, enclosed in a capsule, which for awhile it carries about. The position of the animal, while within the egg, is with the head doubled back towards the tail, one very unfavourable for the process of breathing by internal gills, and hence there is an interesting provision made to meet the emergency. On each side a filament of the substance of the gills projects from the gill-opening, containing vessels in which the blood is exposed to the action of the water. These processes are gradually absorbed after the fish is excluded, until which the internal gills are scarcely capable of respiration. How curious an analogy we here discover with the Frogs and Newts among the Reptiles; and how impressively do we learn the Divine benevolence, when we find that the object of so much contrivance and care is the dreaded and hated Shark !

In these latitudes the Hammer-headed Shark (*Zygæna malleus*), a fish of singular construction, attains a large size. In most particulars it closely

S

resembles the species just noticed, but the head is
widened out on each side into an oblong projection,
at each extremity of which is placed the eye. The
whole of this part has the form of a double-headed
hammer or maul. Undoubtedly one result of this
remarkable structure is a vast increase of the sphere
of vision; but why a fish so formidably armed, and
endowed with such powers of motion, should be thus
favoured, we are not sufficiently acquainted with its
habits to determine.

Another singular deviation from the general structure is found in the Saw-fish (*Pristis antiquorum*),
which is a shark with the head prolonged into a flat

HAMMER-SHARK (*Zygæna malleus*), AND SAW-FISH (*Pristis antiquorum*).

bony sword, each edge of which is armed with sharp
bony spines, resembling teeth, pointing backwards:
there are about twenty of these in each row. The
body also is covered on the upper surface with hard
sharp tubercles, the points of which turn backwards.
In this respect, it resembles some of the Ray or Skate
tribe, as it does also in the flattened form of its body,
and in other respects. Its colour is a dark grey on
the upper parts, gradually softening into white
beneath. This species was known to the ancients,
being found in the Mediterranean Sea, as well as in
the Ocean, but it is in the tropical seas that it acquires
its most gigantic dimensions. It seems to be an animal
of scarcely less ferocity, though far less frequently
met with, than the Common Shark: to the Whales it
is a formidable antagonist, and though the form of its
saw-like sword does not seem most adapted for penc-
trating a resisting body, such is the vigour of its
attack, that it will bury its weapon to the root in the
flesh of the Whale; and instances are not infrequent
in which it has been found firmly imbedded in the
hull of a ship. The following interesting narrative,
by Captain Wilson of the Halifax packet, gives us
an idea of the powers of this monster:—

"Being in the Gulf of Paria, in the ship's cutter,
on the 15th of April, 1839, I fell in with a Spanish
canoe, manned by two men, then in great distress,
who requested me to save their lives and canoe, with
which request I immediately complied; and going
alongside for that purpose, I discovered that they
had got a large Saw-fish entangled in their turtle-
net, which was towing them out to sea, and but for

my assistance they must have lost either their canoe
or their net, or perhaps both, which were their only
means of subsistence. Having only two boys with me
in the boat at the time, I desired them to cut the fish
away, which they refused to do; I then took the
bight of the net from them, and with the joint endeavours of themselves and my boat's crew, we succeeded in hauling up the net, and to our astonishment, after great exertions, we raised the saw of the
fish about eight feet above the surface of the sea.
It was a fortunate circumstance that the fish came
up with the belly towards the boat, or it would have
cut the boat in two.

"I had abandoned all idea of taking the fish, until,
by great good luck, it made towards the land, when
I made another attempt, and having about fifty
fathoms of rope in the boat, we succeeded in making a
running bowline-knot round the saw of the fish, and
this we fortunately made fast on shore. When the
fish found itself secured, it plunged so violently, that
I could not prevail on any one to go near it: the appearance it presented was truly awful. I immediately
went alongside the Lima packet, Capt. Singleton,
and got the assistance of all his ship's crew. By the
time they arrived the fish was rather less violent;
we hauled upon the net again, in which it was still
entangled, and got another fifty fathoms of line made
fast to the saw, and attempted to haul it towards the
shore; but, although mustering thirty hands, we could
not move it an inch. By this time the negroes belonging to Mr. Danglad's estate came flocking to our
assistance, making together with the Spaniards about

one hundred in number: we then hauled on both
ropes for nearly the whole of the day, before the
fish became exhausted. On endeavouring to raise
the fish it became most desperate, sweeping with its
saw from side to side, so that we were compelled to
get strong guy-ropes to prevent it from cutting us to
pieces. After that, one of the Spaniards got on its
back, and at great risk cut through the joint of the
tail, when animation was completely suspended: it
was then measured, and found to be 22 feet long
and 8 feet broad, and weighed nearly 5 tons."*

Other monstrous creatures, of unpleasing forms
and formidable powers, rove at will through these
waters. I shall mention only the Horned Ray
(*Cephaloptera*). Imagine a Thornback or Skate, of
the length of twenty-five feet, with the side-fins
greatly lengthened out, so as to make the total width
upwards of thirty feet: these side-fins, instead of
meeting in a point in front of the head, projecting on
each side into a curved point, like a horn. Such is
the *Cephaloptera;* and it is powerful and voracious
in proportion to its size. Col. Hamilton Smith, in the
neighbourhood of Trinidad, had the pain of witnessing a fellow-creature involved in the horrible embrace
of one of these monsters. It was at early dawn that
a soldier was endeavouring to desert from the ship by
swimming on shore. A sailor from aloft, seeing the
approach of one of these terrific fishes, alarmed the
swimmer, who endeavoured to return; but, in sight
of his comrades, was presently overtaken, the creature throwing over him one of its huge fins, and thus

* Mag. Nat. Hist. 1839, p. 519.

carrying him down. In the following record, which
was inserted in a late Barbadoes paper, though the
description is not drawn up exactly as a Naturalist
would have done it, one has no difficulty in recognis-
ing an enormous *Cephaloptera:*—"On the 22nd of
August [1843], the Brig Rowena was lying in La
Guayra Roads, the weather perfectly calm: I discovered the vessel moving about among the shipping.
I could not conceive what could be the matter. I
gave orders to heave in, and see if the anchor was
gone, but it was not: but to my surprise, I found a
tremendous monster entangled fast in the buoy-rope,
and moving the anchor slowly along the bottom. I
then had the fish towed on shore. It was of a flat-
tish shape, something like a *devil-fish*, but very
curious shape, being wider than it was long, and
having two tusks, one on each side of the mouth,
and a very small tail in proportion to the fish, and
exactly like a bat's tail. The tail can be seen on
board the Brig Rowena. Dimensions of the fish
were as follows:—length from end of tail to end of
tusks, 18 feet; from wing to wing, 20 feet; the
mouth, 4 feet wide; and its weight, 3502 lbs."

Every one may imagine how much the tedium of
a long voyage is relieved by the company of other
vessels, or even by the speaking of a passing ship;
but a few who have only seen vessels lying in tiers,
side by side, at quays, or wharfs, are at all aware of,
or can readily understand, the anxious care with
which commanders guard against two ships on the
high sea coming within even a considerable distance
of each other. I have often been amused by hearing

the wishes expressed by passengers on their first
voyage, when a vessel is speaking at what they think
a most uncivil distance, that she would but come
nearer, particularly if the wind is light, as "there
can be no danger then." Little do they think that
when in a perfect calm the danger of contact is even
greatest, as, if there be wind enough to give the vessel "steerage way," she is under control, and the
evil may be avoided. On this subject, and on the
motions of ships in calms, an unexceptionable authority, Captain Basil Hall, thus speaks :—

"How it happens I do not know, but on occasions
of perfect calm, or such as appear to be perfect calm,
the ships of a fleet generally drift away from one
another, so that, at the end of a few hours, the whole
circle bounded by the horizon is speckled over with
these unmanageable hulks, as they may for the time
be considered. It will occasionally happen, indeed,
that two ships draw so near in a calm as to incur
some risk of falling on board one another. I need
scarcely mention that even in the smoothest water
ever found in the open sea, two large ships coming
into actual contact must prove a formidable encounter.
As long as they are apart, their gentle and rather
graceful movements are fit subjects of admiration;
and I have often seen people gaze for an hour at a
time at the ships of a becalmed fleet, slowly twisting
round, changing their position, and rolling from side
to side as silently as if they had been in harbour, or
accompanied only by the faint rippling sound trip-
ping along the water-line, as the copper below the
bends alternately sunk into the sea, or rose out of it,

dripping wet, and shining as bright and clean as a new coin, from the constant friction of the Ocean during the previous rapid passage across the Trade-winds.

"But all this picturesque admiration changes to alarm when ships come so close as to risk a contact; for these motions, which appear so slow and gentle to the eye, are irresistible in their force; and as the chances are against the two vessels moving exactly in the same direction at the same moment, they must speedily grind or tear one another to pieces. Supposing them to come in contact side by side, the first roll would probably tear away the fore and main channels of both ships; the next roll, by interlacing the lower yards, and entangling the spars of one ship with the shrouds and backstays of the other, would, in all likelihood, bring down all three masts of both ships, not piecemeal, as the poet hath it, but in one furious crash. Beneath the ruins of the spars, the coils of rigging, and the enormous folds of canvas, might lie crushed many of the best hands, who, from being always the foremost to spring forward in such seasons of danger, are surest to be sacrificed. After this first catastrophe, the ships would probably drift away from one another for a little while, only to tumble together again and again, till they had ground one another to the water's edge, and one, or both of them, would fill, and go down. In such encounters it is impossible to stop the mischief; and oak and iron break and crumble in pieces like sealing-wax and pie-crust. Many instances of such accidents are on record, but I never witnessed one.

"To prevent these frightful rencontres, care is always taken to hoist out the boats in good time, if need be, to tow the ships apart, or, what is generally sufficient, to tow the ships' heads in opposite directions. I scarcely know why this should have the effect; but certainly it appears that, be the calm ever so complete, or *dead*, as the term is, a vessel generally *forges ahead*, or steals along imperceptibly in the direction she is looking to; possibly from the conformation of the hull."*

But there are indications of our patience being at length rewarded by a breeze from the eastward; and now it comes, rippling the surface as it approaches, turning that into a deep uniform blue which has so long borne a glassy brightness reflected from the sky. The seamen are joyous and alert, for they know that this is no "cat's-paw," but the "regular trade." Now it strikes the ship; the sails, gracefully swelling, receive the unwonted impulse; and the lengthened wake, where the water coils and frets in the newly-cut furrow, tells that the vessel makes way once more. The breeze freshens; the little waves become larger, and, arching over each other, break with patches of whitening foam; every sail is speedily set that will draw; and we run gaily along towards the west, under an eight-knot breeze. We can scarcely stop to notice the amity that subsists between the Shark and the Pilot-fish (*Naucrates ductor*), a beautiful little creature, about the size of a herring, the back striped transversely with broad alternate bands of brown and

* Frag. Voy. and Trav. 2nd Series, i. p. 226.

bright azure; nor the three or four pretty little
Rudder-fishes (*Perca saltatrix*, LINN.), which have
been following and accompanying us for several
days past. These are amusing little creatures. They
are about six inches long, yellowish brown, with
pale spots: they keep close to the stern, in the angle
formed by the rudder and the counter of the ship,
the "dead water," as it is called by seamen. Hence
they occasionally dart out after any little atom of
floating or sinking substance which promises to be
eatable, and then, having either seized or rejected it,
scuttle back again to their corner, remaining there
day and night without rest. Nor can we do more
than glance at the Sucking-fishes (*Echeneis*), that
are swimming around, or have attached themselves
to the side of the rudder by means of the singular
oval disk on the head. As this organ is of singular
construction, so its use in the economy of the animal
is involved in entire obscurity. The theory of the
fish being a very slow swimmer, and needing to be
carried along by others, must have been formed by
persons who never had an opportunity of seeing the
Remora alive. I have seen many, and could detect
no inferiority in their powers of swimming to a
young Shark of the same size, which they much resemble in general appearance and motion, when in
the water. There seems to be a perfect vacuum
formed by the adhesion of the disk, and the external
pressure, when under water, is of course great. As
the mouth opens upon the upper surface of the muzzle, owing to the projection of the lower jaw, it
is possible that this habit may be connected with

taking food: there are many little creatures, such
as *Crustacea*, Barnacles, &c., that are parasitical on
the bodies of marine animals, or attach themselves
to any submerged substance. If the *Echeneis* feeds
on these, there is an obvious reason why the head
should be affixed to the surface during the dislodg-
ment of the adhering prey, in order to acquire
greater steadiness, as well as a leverage by which
to act more effectively. At all events, we know
that it is not a useless habit; we trace enough
of manifest design and contrivance in what we do
know of the animal creation, to warrant our con-
fident conclusion, when we find any instinct, the
intention of which is not obvious, that it also is
the production of infinite wisdom and goodness,
and that it could not have been spared without
injury to the animal.

Borne on the wings of the welcome breeze, we
rapidly approach that archipelago of lovely islands
that gladdened the heart and rewarded the zeal of
the chivalric WORLD-FINDER, the first fruits of the
vast continent which the genius and daring of one
master-mind opened to astonished Europe. The
joyful sound of "Land in sight!" resounds through
the ship, and yonder, upon the bow, is discovered,
rising out of the blue sea, the beautiful island of
Antigua. As we draw near, we are struck with its
loveliness; the coast is low, but the land rises behind
into rounded hills of moderate elevation, whose
swelling eminences and gentle slopes assume somewhat of the appearance of the chalk hills and downs
of our own sweet England. But there are features

which effectually distinguish this island from our own, and fail not to remind us that we are beholding the gorgeousness of the tropics. The summits of the hills are clothed with magnificent forest-trees of strange forms and foliage; the graceful palms wave their feathery crowns against the deep blue sky: leafless cacti, thick and cylindrical, project from the rocks, or take the shape of enormous candelabra: the great American aloe, with its thick and spiny leaves, shoots up its glorious head of yellow blossoms to the height of twenty feet: the clusters of golden fruit depend from the plantain and banana, whose gigantic fronds are cut by the winds into ragged segments; while the whole array is bound and matted together by strong rope-like climbing plants, which, crossing each other in every direction, and twisting around the forest-trees, and around each other, like huge cables, present an immense net of vegetation, impenetrable except by the axe of the woodman. Tree-ferns, possessing all the grace and elegance of those with which we are familiar, but growing to a giant size, shoot up from the clefts of the rocks, or from the branches of the loftier trees, their rich brown stalks contrasting with the vivid green of their fan-shaped fronds. The sides of the hills are clothed with luxuriant plantations of Indian corn, or the still more rich and beautiful sugar-cane; and here and there a walk of cocoa-trees is rendered conspicuous by the glowing scarlet blossoms of the coral trees, by whose shadow they are sheltered from the vertical sun. The coast is broken into numerous little bays and coves, some penetrating far into the island, like

canals among the plantations. A multitude of little
islets are scattered around on the surface of the sea,
on many of which the cattle are grazing on the rich
and succulent pasture. Some of them, however,
are little more than accumulations of sand, formed
of powdered coral and sea-shells, and affording support only to some coarse sedges, and to mangrove-
trees. The latter, indeed, delights in such situations, flourishing at the very edge of the sea, and
even where the ground is continually liable to
inundation. The contorted roots of this tree grow
to a considerable extent above the soil, so that the
base of the trunk is elevated on a cone of matted
roots, through which the water washes, while from
the branches young twigs are perpetually shooting
downward, till, reaching the soil, they take root,
and send forth other shoots: thus, in a few years,
a single plant will spread into a grove, and cover
a large space of land. As we sail with tortuous
course through these delightful groups of ever-
verdant isles, fresh scenes of beauty are continually
rising before us. Now a conical hill, of regular
form, arrests the attention, clothed with thick foliage
from the water's edge to the summit, where the white
clouds appear to rest: then we admire the irregular
surface of another isle, whose dark ravines seem
to acquire additional gloom from the glowing sun-
light that plays upon the surrounding eminences:
here a little islet of bright green looks in the blue
sea like an emerald set in sapphire; there the bold
cliffs and black precipices of a larger island an-
nounce a very different formation. Now and then

T

we open a small but deep and beautiful bay. "A
pretty little village or plantation appears at the
bottom of the cove: the sandy beach stretches like
a line of silver round the blue water, and the cane-
fields form a broad belt of vivid green in the back-
ground. Behind this, the mountains rise in the
most fantastic shapes, here cloven into deep chasms,
there darting into arrowy points, and every where
shrouded, and swathed, as it were, in wood, which
the hand of man will probably never lay low. The
clouds, which within the tropics are infallibly attracted by any woody eminences, contribute greatly
to the wildness of the scene: sometimes they are
so dense as to bury the mountains in darkness, at
other times they float transparently like a silken
veil; frequently the flaws from the gulleys perforate
the vapours, and make windows in the smoky mass;
and then, again, the wind and the sun will cause the
whole to be drawn upwards majestically, like the
curtain of a gorgeous theatre."

Around these islands the water is frequently shallow, a fact made sufficiently obvious by its colour:
instead of the deep-blue tint which marks the un-
fathomed Ocean, the water on these shoals becomes
of a bright pea-green, caused by the nearness of
the yellow sands at the bottom; and the shallower
the water, the paler is the tint. The light thrown
upwards by reflection upon the under part of the
swollen sails, transfers the same hue to them, giving
them a singular aspect; but once I observed a still
more curious appearance, arising from the same
cause. Being becalmed off one of the little Keys

of the Florida Reef, the crew had been amusing themselves with fishing, in which they had been very successful. An Osprey (*Haliæetus ossifragus*), attracted, doubtless, by the fish that lay in profusion about the decks, was slowly sailing around, occasionally alighting on the ropes and spars. As he hovered overhead, turning his head from side to side, every feather was distinctly seen; but from the reflection of the water beneath, all his under parts, which are pure white, appeared of a fine pea-green, and it was only on catching a side-glance at him, that I discovered his true colour, and identified the species. It is very pleasing to peer down into the varying depths, especially in the clear waters of these seas, and look at the many-coloured bottom; sometimes a bright pearly sand, spotted with shells and corals; then a large patch of brown rock, whose gaping clefts and fissures are but half hidden by the waving tangles of purple weed; where multitudes of strange creatures revel and riot undisturbed.

"Come down, come down from the tall ship's side;
    What a marvellous sight is here!
Look! purple rocks and crimson trees,
    Down in the deep so clear!

"See! where those shoals of dolphins go,
    A glad and glorious band:
Sporting amidst the day-bright woods
    Of a coral fairy land.

"See! on the violet sands beneath,
    How the gorgeous shells do glide!
O sea! old sea! who yet knows half
    Of thy wonders and thy pride?

" Look, how the sea-plants trembling float,
   All like a mermaid's locks,
Waving in thread of ruby red,
   Over those nether rocks !

" Heaving and sinking, soft and fair,
   Here hyacinth—there green,—
With many a stem of golden growth,
   And starry flowers between.

" But away ! away ! to upper day !
   For monstrous shapes are here ;
Monsters of dark and wallowing bulk,
   And horny eyeballs drear :

" The tusked mouth and the spiny fin,
   Speckled and warted back,
The glittering swift and flabby slow,
   Ramp through this deep sea track.

" Away ! away ! to upper day !
   To glance o'er the breezy brine,
And see the nautilus gladly sail,
   The flying-fish leap and shine !"

While pursuing our pleasant course amidst these
sandy keys, we may often observe the Green Turtle
(*Chelonia mydas*) swimming or floating at the surface. In general it is difficult to approach them
within less than a few yards, as they are very wary,
and dive with great rapidity. The shoals and reefs
surrounding the islands, where the sun penetrates
and warms the water, are favourite resorts of these
marine *Reptilia;* and here, too, grow in abundance
the sea-plants (*Zostera*, &c.) on which they feed.
At night, the females land on the low sandy beaches,
and after examining the place with great caution
and circumspection, lay their eggs in holes, which
they scoop out with their fin-like feet. The work

being accomplished, the sand is again scraped back over the eggs, and the surface made smooth as before. The sun soon hatches the eggs, and the little Turtles crawling forth from the sand, betake themselves to the sea. The usefulness of this animal as an article of luxurious food is well known; but its real value can only be appreciated, when we view it as affording an immediate relief from the horrors of scurvy, which, arising from the constant use of salted provisions, has often proved so terrible a scourge in long voyages. There is a peculiarity in the structure of the heart of this and kindred animals, which is worthy of notice. In man and other warm-blooded animals, the blood is brought by the veins to the heart, and poured into a chamber called the *right auricle;* a communication exists between this and a second chamber, called the *right ventricle;* from the latter the blood is forced through a large artery to the lungs, to be renewed by exposure to the air; from the lungs it is sent through veins to a third chamber of the heart, called the *left auricle,* and thence into a fourth, called the *left ventricle,* from which the great artery, called the *aorta,* carries it again into the whole body. Thus, no particle of the blood can be conveyed again into the system without having passed through the lungs; but in the Turtle the case is different. All the four chambers of the heart are present, but there is a communication open between the *left* and *right ventricles;* and the *aorta* and *pulmonary artery* both originate from the *right ventricle.* In consequence, a part only of the blood is sent thence to the lungs, which,

returning through the *left auricle* and *ventricle*, is
thrown into the *right ventricle*, and mixed with that
which is just brought from the body; the mixed
blood being partly returned to the body through the
aorta, and partly sent to the lungs. But this is the
course only when the animal is breathing; and as a
large part of its life is passed under water, this contrivance enables the circulation to go on under circumstances when breathing necessarily ceases. For
if no air enters the lungs, the blood cannot pass
through them; therefore, when under water, the
blood passing through the *right auricle* and *ventricle*,
is immediately sent by the *aorta* into the body without any exposure to the air. Of course, as the blood
thus unrenewed would become more and more impure, this could not proceed very long without loss
of life, and hence there is a limit to the period
during which the breathing may be suspended, when
the animal must come to the surface or die.

Many of the fishes of these seas partake of the
brilliancy of colour with which the birds and insects
of the same sunny region are so lavishly adorned.
I have seen some of great beauty readily captured
with a hook from the deck of a vessel in shallow
water;—such as the Yellow-fin (*Sparus synagris*,
LINN.), which has its body marked with longitudinal
bands of delicate pink and yellow alternately; the
fins are bright yellow, and the tail fine pale crimson.
A larger species, which the seamen denominated the
Market-fish (*Labrus anthias*, L.), is all over of a
silvery tint with a ruddy glow, the fins and tail
bright crimson; this species has very large scales.

Then there is the Hog-fish (*Labrus flavus*, L.?), of singular beauty, shaped somewhat like a perch, with silvery grey scales; the head marked all over with streaks of brilliant violet blue, fantastically arranged, somewhat like the stripes upon the head of the Zebra. Still, however, even here there is some deformity; at least, every thing does not accord with our habitual ideas of comeliness; these beauties are set off, as by a foil, by the visage of the Cat-fish (*Silurus catus*), a creature of remarkably hideous aspect, but which is esteemed as food.

In some of the quiet nooks and sheltered bays of these lovely islands, where the vegetation is green and luxuriant to the water's edge, we may catch a sight of a herd of Manatees, or Sea-Cows. These animals are usually classed with the Whales, but they seem, indeed, to be much more intimately connected with the *Pachydermata*, an order that contains the Elephant and Hippopotamus. The form is long and tapering, but plump, and has been compared to that of a filled wine-skin or leather bottle. The hinder feet are altogether wanting, but the fore limbs assume the appearance of broad flat fins or flippers, the fingers of which are not separated externally, but can be distinctly felt through the skin; and the nails or claws by which the paw is terminated, sufficiently indicate their presence. These creatures are perfectly inoffensive in their manners, timid, and retiring; they delight in secluded places, shallow creeks, and particularly the mouths of the great South American rivers, often proceeding many miles up the country. For such situations they are

peculiarly adapted; the broad valleys of these regions, parched up to barrenness in the dry season,
and then inundated, so as to resemble seas during
the periodical rains, would not be suited to the capa-
cities of a terrestrial ruminant; but the aquatic
habits of the Manatee enable it to avail itself of the
rich and abundant vegetation of the watery expanse,
as well as to range the coast when it is parched up
by the returning drought. Being exclusively her-
bivorous, the flesh is highly esteemed; its flavour is
thought to resemble that of excellent pork, though
by some it has been rather compared to beef. Hunt-
ing this animal is a favourite amusement in the
countries of its resort; a party proceed in a small
boat to its haunt, furnished with a harpoon, to
which is attached a stout line; when the weapon
is infixed, the creature dives; in the meanwhile the
boat is rowed ashore, and the Manatee, exhausted
by its efforts to escape, is drawn on land by the
cord, and despatched. Many of its habits are exceedingly interesting: it is fond of sporting in the
water, and leaping from the surface in the manner
of the true *Cetacea.* Such is the attachment evinced
by these animals for each other, that it is said, when
one is harpooned, the rest of the herd will assemble,
and endeavour to drag out the harpoon with their
teeth. When basking on the shore, the young are
collected into the centre of the group for protection, and if a calf has been killed, the mother will
suffer herself to be secured without effort; while,
on the other hand, if the dam be taken, the young
will follow the boat to the shore.

THE PACIFIC OCEAN.

WHEN the astonishing sagacity and enterprise of the Genoese had discovered the confines of a new world across the trackless Atlantic, it was without hesitation concluded, not only by himself, but by all Europe, that the new land formed the extreme eastern shore of Asia; and hence the name of Indies, by this mistake, was given to these islands, which has been perpetuated even to the present time. Aware of the round form of the earth, the geographers of that age could well conceive the possibility of reaching India by a westerly course; but, ignorant of the magnitude of the globe, they had formed a very inadequate idea of its existence, being totally unaware of the vast continent, and still vaster ocean, which separated Asia from the Atlantic. But as, impelled by an insatiable thirst for gold, the unprincipled Spaniards pushed their career of robbery and murder farther and farther into the continent, they began to hear tidings of a boundless sea, which stretched away to the south and west, beyond the horizon of the setting sun. Balboa, one of the reckless spirits who sought fortune and fame at all hazards in the newly-found regions, boldly determined to seek the sea of which the Indians spake. At the head of a little band of men, guided by a Mexican,

he succeeded, after severe privations and imminent
dangers, in crossing the isthmus that connects the
northern and southern portions of the continent.
They had arrived at the foot of a hill, from the top
of which the Indian assured him he would obtain a
sight of the wished-for sea; when in the enthusiasm

BALBOA DISCOVERS THE PACIFIC.

of the moment, leaving his companions behind, the
Spanish chief ran to the summit, and beheld a limit-
less Ocean sleeping in its immensity at his feet.
With the spurious piety common to the times—a
piety that could consist with the grossest injustice,
the blackest perjury, and the most barbarous cruelty,
—he knelt down and gave thanks aloud to God for
such a termination of his toils; then having descend-

ed the cliffs to the shore of the Ocean, he bathed in its mighty waters, taking possession of it by the name of the Great South Sea, on behalf of the King of Spain. This was in the year 1513; but it was not till seven years afterwards that its surface was ruffled by a European keel. Then Magalhaens or Magellan, a Portuguese navigator of great ability, in the service of Spain, having run down the coast of South America, discovered the straits which have since borne his name, through which he sailed, and emerging from them on the 28th November, 1520, first launched out upon the broad bosom of the South Sea. For three months and twenty days he sailed across it, during which long period its surface was never ruffled by a storm; and from this circumstance he gave to the Ocean the appellation of the Pacific, which it still retains. The immediate vicinity of the Straits, however, has been considered peculiarly subject to tempests; while the almost continual prevalence of westerly winds, joined to the severity of the climate, has always given a character of difficulty and hazard to the passage from the one Ocean to the other.

In approaching the extreme point of South America, navigators have been struck with the extraordinary size of a floating sea-weed, the *Macrocystes pyrifera* of botanists. It consists of a smooth round stem, commonly from 500 to 1000 feet in length: Foster mentions one which was 800 feet, and some specimens are reported even to attain the enormous dimensions of 1500 feet. From the stem grow a great number of pear-shaped air-vessels, which end

in long, flat, wrinkled fronds of a semi-transparent
brown hue. I have already spoken of the Gulf-weed
(*Sargassum vulgare*), as being met with in particular
parts of the Altantic: similar collections of it occur
also in these and other seas, and much mystery
seems to lie about its origin and mode of growth.
From specimens having been found with roots, it
appears certain that in a living state it is attached
to the bottom, whence it is not impossible that it
may be detached spontaneously at a certain period
of its growth, that the seed-vessels may be perfected
by exposure to light and air. Near the shores sea-
weeds are found so uniformly growing to rocks as
to form a very valuable indication of the presence
of hidden dangers. These appear to be chiefly of
the former kind.

To these remote and inhospitable seas many vessels are annually despatched from this country, as
well as from the United States, in pursuit of various
species of Seals, and of the Sperm Whale. To obtain
the former, they resort to any of the small islands
which are scattered over the southern part of the
Atlantic and Pacific, but particularly those which lie
around Cape Horn. These animals yield two valuable products, oil and fur; but not indiscriminately,
the oil being afforded by the Elephant Seal (*Macro-
rhinus proboscideus*), a singular animal, of large size;
being often seen thirty feet long, and eighteen round
at the thickest part. A very remarkable formation
of the snout has given the distinctive name to this
species. At a certain season of the year, in the
adult males, the skin of the tip of the nose, which

covers a number of cells ordinarily empty, becomes enlarged and lengthened by the blood that the animal has the power of forcing into the cells. This projection is now a foot in length; but it appears to be nothing more than a mere appendage, somewhat resembling, in more respects than one, the fleshy wattle on the head of the turkey, which can be similarly inflated. In the spring—that is, in these latitudes, the months of August and September— the Elephant Seals betake themselves to the rocky shores in large herds: at this time they are exceedingly fat, and a single male will sometimes yield a butt of oil. They remain on shore until the middle of summer, when the young, which have been born in the mean time, are fit to take the water and provide for themselves. As the old ones have taken no food during the whole of this period, they are become very lean and weak, but soon recruit their powers. Though furnished with large and powerful tusks, and endowed with sufficient strength to use them, the Sea-Elephant is a most mild and inoffensive creature, suffering the seamen not only to walk among them uninjured, but even to bathe in the midst of the herd when swimming, with perfect impunity. In self-defence, however, or in defence of their young, their resistance becomes formidable. One of Anson's men, having killed a young one, had the cruelty and rashness to skin it in the presence of its mother: but she, coming behind him, got the sailor's head into her mouth, and so scored and notched his skull with her sharp teeth, that he died in a day or two afterwards.

U

Among themselves, however, the males are accustomed to fight at certain periods with great ferocity. "Their mode of battle is very singular. The two rival giant knights waddle heavily along; they meet and join snout to snout; they then raise the

ELEPHANT SEALS, FIGHTING.

fore part of the body as far as the fore paws, and open their immense mouths; their eyes are inflamed with rage, and they dash against each other with the greatest violence in their power: now they tumble one over the other; teeth crash with teeth, and jaws with jaws; they wound each other deeply, sometimes knocking out each other's eyes, and more frequently their tusks; the blood flows abundantly; but these raging foes, without ever seeming to observe it, prosecute the combat till their strength is completely exhausted. It is seldom that either is left dead on the field, and the wounds they inflict, however deep, heal with inconceivable rapidity. The object of these encounters is to obtain the lordship of a herd of females, by which a male is always accompanied, and over which he rules with undivided empire."

While on land, the motions of these animals are slow and unwieldy, and apparently productive of much fatigue. Their gait is described as singular: as they crawl along, the vast body trembles like a great bag of jelly, owing to the mass of blubber by which the whole animal is invested, and which is as thick as it is in a whale. After having proceeded thus for fifteen or twenty yards, they halt to rest; and if forced to go forward by repeated blows, their appearance presently manifests the distress to which they are subjected by the increased exertion. It is remarkable that, in these circumstances, the pupil of the eye, which ordinarily is bluish-green, becomes blood-red. They do not, therefore, commonly wander far from the sea, but generally choose low sandy shores, or the mouths of rivers, for their haunts; though they have been known to ascend hills of twenty feet elevation, in search of some pools of water. They appear to be incommoded by the direct beams of the sun ; and, to shelter themselves from its influence, they have the habit of scooping up the wet sand with their forepaws, and throwing it over their bodies, until they are entirely enveloped by it.

It is for the oil which is produced by this species

of Seal that many vessels are sent to the islands of
the Pacific, and to the icy regions of the Antarctic
Ocean.   Its skin, though serviceable as leather for
harness, &c., yields no fur, being clothed only with
coarse hair.   The oil, however, is of a very superior
quality; it is clear and limpid, without any smell,
and never becomes rancid; it burns slowly, and
without smoke or disagreeable odour.   The hunters
destroy the animals with long lances : watching the
instant when the Seal raises the left forepaw to advance, they plunge the lance into its heart, when it
immediately dies.   The fat is then peeled from the
carcass, and cut up and packed in casks in a similar
manner to that of the Whale.

The soft yellow fur, with a changeable gloss, which
a few years ago was so much made into caps, is
another product of a South Sea voyage.   It is the
covering of more than one species of Seal, belonging
to a tribe called Otaries, because their heads are
furnished with external ears, of which the others
are deprived.   That which is by eminence called the
Fur-Seal (*Otaria Falklandica*), is clothed externally
with long hair of a grey hue ; but when this hair is
pulled out, there is seen a thick fur of great soft-
ness, curly or wavy, and of a fine yellowish brown.
The habits of this animal are in general similar to
those of the Sea-Elephant just described: it is, however, much more active on land, often escaping from
a man running.   Its history affords us an instance of
change of instincts produced by experience.   When
the Seals of South Shetland were first visited, they
had no apprehension of danger from man ; but would

unsuspectingly remain while their fellows were slain
and skinned; but latterly they have learned to
guard against the new dangers, by placing themselves on insulated rocks, from which they can in
a moment throw themselves into the water. We
may form a notion of the zeal with which this commercial enterprise was prosecuted, as well as of its
valuable character, if it had been pursued with prudent restrictions, from the fact that in the years
1821 and 1822, there were taken from the South
Shetland Isles, 320,000 skins of Fur-Seals, and 940
tuns of Sea-Elephant oil. The former valuable animal might, by proper precautions, have been made
to produce 100,000 skins annually, for a long time
to come. "This would have followed from not
killing the mothers till the young were able to take
the water; and even then, only those which appeared
to be old, together with a proportion of the males,
thereby diminishing their total number, but in slow
progression. The system of extermination was practised, however, at South Shetland; for whenever a
Seal reached the beach, of whatever denomination,
he was immediately killed and his skin taken; and
by this means, at the end of the second year, the
animals became nearly extinct; the young, having
lost their mothers when only three or four days old,
of course all died, which, at the lowest calculation,
exceeded 100,000."*

Other species of Otaries, which frequent these
seas, have large heads, clothed with long shaggy hair,

* Weddell's Voyage, p. 141.

v 2

which, falling down on the neck, assumes the appearance of a mane, and hence they are frequently called Sea-lions.   Of some of these animals which Captain Cook met with, he says: "It is not at all dangerous to go among them, for they either fled or lay still.   The only danger was in going between them and the sea; for if they took fright at any thing, they would come down in such numbers, that if you could not get out of their way, you would be run over.   When we came suddenly upon them, or waked them out of their sleep (for they are sluggish, sleepy animals), they would raise up their heads, snort and snarl, and look fierce, as if they meant to devour us; but as we advanced upon them, they always ran away, so that they are downright bullies." Like the Sea-Elephant, however, they are quarrelsome among themselves.   They often seize each other with a degree of rage which is not to be described; and many of them are seen with deep gashes on their backs, which they had received in these wars.   Others of the eared Seals are fierce and fearless towards man himself.   Woodes Rogers describes one which he met with at the Galapagos, which he calls a Sea-bear, probably of a species (*Otaria ursina*) common in the seas of which I am speaking. He says, "A very large one made at me three several times; and if I had not happened to have a pikestaff headed with iron, he might have killed me.   I was on the level sand when he came open-mouthed at me from the water, as fierce and quick as an angry dog let loose.   All the three times he made at me, I struck the pike into his breast,

which at last forced him to retire into the water,
snarling with an ugly noise, and showing his long
teeth."*

Dividing the dominion of these inhospitable islands
with the Seals, may be seen myriads of Penguins;
curious birds, which seem to be the link which connects the feathered with the finny race. Their
little wings, destitute of quills, but covered with stiff
scaly feathers, hang down by their sides, perfectly
incompetent to lift them from the ground, resembling in shape the fins of a fish, or still more the
flippers of a turtle. But see the Penguin in the
water; the deficiency of flight is abundantly compensated by the power and agility it possesses in
this element: it dashes along over the surface in
gallant style, or diving, shoots through the water
with the rapidity of a fish, urging its course by the
united action of its finny wings and its broad
webbed feet; then, coming again to the top, leaps
over any obstacle in its course, many feet at a bound,
and pursues its way. On the sandy shores or flat
rocks in the Southern Ocean, the Penguins, of several
species, assemble in innumerable multitudes, for the
purpose of hatching their eggs and rearing their
young. The feet are placed very far back on the
body, so that the bird assumes an erect position when
resting or walking on land; and from their posture,
their colours, their numbers, and their orderly arrangement, they have been compared, when seen at a
distance, to an army of disciplined soldiers. One voyager likens them to a troop of little children standing

* Kerr's Voyages, x. 374.

up in white aprons, from their white bellies contrasting with their blue backs. The presence of these birds is described as greatly increasing the dreary character of these desolate regions; their perfect indifference to man conveying an almost awful impression of their loneliness. The intrusion of seamen even into the very midst of them causes no alarm; no resistance is offered, no escape is attempted; the birds immediately gaze around with a sidelong glance at the visitors, but they move not from their eggs, standing quietly while their companions are one by one knocked on the head, and waiting without dread till their own turn comes. We can scarcely form an adequate idea of one of these camps or towns, as they have been appropriately called. A space of ground, covering three or four acres, is laid out and levelled, and then divided into squares for the nests, as accurately as if done by a surveyor: between these compartments they march and countermarch with an order and regularity that remind one of soldiers on parade. But what shall we say to a colony of these birds, the King Penguin (*Aptenodytes patachonica*), which was seen by Mr. G. Bennett, on Macquarie Island? It covered thirty or forty acres; and though no conjecture could possibly be formed of the number of birds composing the town, yet some notion of its amazing amount may be given from the fact, that during the whole day and night 30,000 or 40,000 are continually landing, and as many going to sea. There are three principal species, which inhabit the southern portion of the globe, which bear great resemblance to each other

in manners, and generally are found in company.
These are the one just mentioned, the Crested Penguin (*A. chrysocome*), and the Jackass Penguin (*A.
demersa*). The latter has obtained its title from its
nightly habit of emitting discordant sounds, which

PENGUINS.

have been likened to the effusions of our humble
sonorous friend of the common. This species seems
to deviate from the general manner of breeding, as
it burrows on the sandy hills, and is more sensible
of injury than its fellows. For Forster describes the
ground as every where so much bored, that a person

in walking often sinks up to the knees; and if the
Penguin chance to be in her hole, she revenges herself on the passenger by fastening on his legs, which
she bites very hard.

The following notices of these singular birds, by
those who have seen them in their haunts, are interesting, as illustrative of their economy:—"One day,"
says Mr. Darwin, "having placed myself between
a Penguin and the water, I was much amused by
watching its habits. It was a brave bird; and, till
reaching the sea, it regularly fought and drove me
backwards. Nothing less than heavy blows would
have stopped him; every inch gained he firmly kept,
standing close before me, erect and determined.
When thus opposed, he continually rolled his head
from side to side in a very odd manner, as if the
power of vision lay only in the anterior and basal
part of each eye. This bird is commonly called the
Jackass Penguin, from its habit, while on shore, of
throwing its head backwards, and making a loud
strange noise, very like the braying of that animal;
but while at sea and undisturbed, its note is very
deep and solemn, and is often heard in the night-
time. In diving, its little plumeless wings are used
as fins; but on the land, as front legs. When crawl-.
ing (it may be said on four legs) through the tussocks, or on the side of a grassy cliff, it moved so
very quickly that it might readily have been mistaken for a quadruped. When at sea and fishing,
it comes to the surface for the purpose of breathing,
with such a spring, and dives again so instantaneously, that I defy any one at first sight to be sure

that it is not a fish leaping for sport."* Of the same
species, apparently, Captain Fitzroy thus speaks:—
"Multitudes of Penguins were swarming together
in some parts of the island [Noir Island], among the
bushes and tussocks near the shore, having gone
there for the purposes of moulting and rearing their
young. They were very valiant in self-defence, and
ran, open-mouthed, by dozens, at any one who in-
vaded their territory, little knowing how soon a stick
could scatter them on the ground. The young were
good eating, but the others proved to be black and
tough when cooked. The manner in which they
feed their young is curious, and rather amusing:
the old bird gets on a little eminence, and makes
a great noise, between quacking and braying, hold
ing its head up in the air, as if it were haranguing
the penguinnery, while the young one stands close to
it, but a little lower. The old bird having continued
its clatter for about a minute, puts its head down,
and opens its mouth widely, into which the young
one thrusts its head, and then appears to suck from
the throat of its mother for a minute or two, after
which the clatter is repeated, and the young one
is again fed; this continues for about ten minutes.
I observed some that were moulting make the same
noise, and then apparently swallow what they thus
supplied themselves with; so, in this way, I suppose,
they are furnished with subsistence during the time
they cannot seek it in the water."† Mr. Weddell
observes of the King Penguins:—"In pride these

* Voyages of Adventure and Beagle, iii. 256.   † Ibid. i. 387.

birds are perhaps not surpassed even by the peacock, to which, in beauty of plumage, they are indeed
very little inferior. During the time of moulting,
they seem to repel each other with disgust, on
account of the ragged state of their coats; but as they
arrive at the maximum of splendour, they re-assem-
ble, and no one who has not completed his plumage
is allowed to enter the community. Their frequently
looking down their front and sides, in order to con-.
template the perfection of their exterior brilliancy,
and to remove any speck which might sully it, is
truly amusing to an observer.

"About the beginning of January they pair and
lay their eggs. During the time of hatching, the
male is remarkably assiduous, so that when the hen
has occasion to go off to feed and wash, the egg is
transported to him; which is done by placing their
toes together, and rolling it from the one to the
other, using their beaks to place it properly. As
they have no nest, it is to be remarked that the egg
is carried between the tail and legs, where the female,
in particular, has a cavity for the purpose.

"The hen keeps charge of her young nearly a
twelvemonth, during which time they change and
complete their plumage; and in teaching them to
swim, the mother has frequently to use some artifice; for when the young one refuses to take the
water, she entices it to the side of a rock and cun-
ningly pushes it in; and this is repeated until it takes
the sea of its own accord."* All the species are
arrant thieves, each losing no opportunity of stealing

* Voyage towards the South Pole, p. 55.

materials during nest-building time, and even the eggs from each other, if they are left unguarded. They are usually thought, when seen at sea, to indicate that land is at no great distance; but this indication is not always correct, for they are occasionally seen very far from any shore, and, indeed, with their swimming powers, one can readily imagine that the space of a few leagues would be no object of concern. The Crested Penguin, in particular, lives in open sea; it has been seen some hundreds of miles from land, voyaging in pairs, male and female.

The chief object of commercial speculation in the Pacific is the pursuit of the Sperm Whale, than which the whole wide range of human enterprise affords no occupation of more daring adventure, or more romantic interest. A crew of thirty or forty hardy fellows leave their native land, and boldly steer away to the most distant parts of the globe. The tempestuous sea of Cape Horn soon finds them hotly engaged in striking their giant game; or, if they find it not here, they do not hesitate to stretch away to the shores of New Zealand, or even to seek the leviathan of the deep five thousand miles farther, in the distant seas of China and Japan. Now they are braving the horrors of the Antarctic sea, threading an intricate and perilous course through fields and bergs of floating ice, "under the frozen serpent of the south;" anon they are upon the equator, toiling with undaunted spirit beneath the rays of a vertical sun. The bleak and barren rocks of the Horn, tenanted by Penguins, are for-

saken for the sunny isles of Polynesia, and these,
again, for the inhospitable shores of Kamschatka.
Peculiar dangers attend them in their protracted
voyage; if they escape unscathed from the storms
of the south, it is to enter an ocean strewn with innumerable reefs of stony coral, whose positions are
but imperfectly indicated in charts, to touch one of
which would be inevitable destruction; if these are
safely passed, it is to penetrate into a sea vexed with
the most terrible of tempests, the typhoon. The
duration of the voyage is protracted to a length
which would justify our calling it an exile; this is
no summer's trip; three and even four years are
the ordinary periods allotted to this enterprise. The
object of the pursuit, gigantic in size and power,
seems to demand no ordinary courage in its assail-
ant; and more especially in his own element, when
he is "making the sea to boil like a pot of oint-
ment," to venture to the battle in a frail boat, needs
a hardihood of more than common calibre. The
moment of victory is frequently the moment of
danger; the dying struggles of the lanced Whale
are of fearful impetuosity; the huge and muscular
tail lashes the Ocean into foam, and the long and
powerful lower jaw, serried with teeth, snaps convulsively in every direction. Timid as this mighty
animal usually is, instances are not infrequent, in
which a consciousness of strength has been accompanied by the will to use it. The destruction of
the ship Essex, an American whaler, affords a remarkable instance of the ferocity and determination,
as well as of the power, of the Sperm Whale. This

vessel was whaling in the vicinity of the Society
Islands, when one of these animals, having grazed
its back in passing beneath the vessel's keel, became
enraged, and after swimming to some distance, suddenly turned, and rushed with amazing force against
the ship. The helmsman vainly endeavoured to
avoid the blow, and the animal, repeating the attack,
stove in the ship's bows, when she speedily filled
and went down, barely allowing the hands on board
time to take to the boat. Those who were out in
pursuit, seeing, to their astonishment, their vessel
sink without any apparent cause, hastened to the
spot, and the whole crew found themselves in open
boats, three thousand miles from the coast of Chili,
to which they determined to proceed, but where
three or four only arrived, after painful and protracted sufferings.

The Sperm Whale (*Physeter macrocephalus*) attains
a greater length than the Greenland Whale, from
which it is at once distinguished by the remarkable
form of the head. As in the latter, the head occupies about one-third of the entire length, but it
is of the same thickness throughout, appearing as
if it had been suddenly cut off at the muzzle; so
that the head bears no small resemblance to a huge
box. There is no whalebone; but the lower jaw,
which is narrow, and fits into the upper, is armed
with a series of sharp teeth, which are received into
hollows in the upper gums. The blow-hole is placed
at the front angle of the head; the eye is just above
the inner corner of the mouth, and over this, where
the head joins the body, there is a hunch, called the

bunch of the neck; from hence the body is nearly
straight to within one-third of its length from the
tail, where there is a larger prominence called the
hump; it now rapidly tapers away to the tail: the
whalers distinguish this tapering part by the name
of "the small," and the broad horizontal tail, as "the
flukes." The whole of the upper portion of the
square and bluff head is occupied by a cavity, tech-
nically termed "the case;" which is not covered by
bone, but by a thick, tendinous, elastic skin, and
lined with a beautiful glistening membrane. This
cavity is filled with a clear oil, which, after death,
cools into the substance well known as spermaceti.
Some idea may be formed of the capacity of the case,
from the fact that, in a large Whale, it will frequently
be found to contain ten large barrels of this valuable
product. Immediately beneath the case is placed
"the junk," a thick triangular mass of tough elastic
substance, which also yields a considerable quantity
of spermaceti. The fins are comparatively small, and
are situated a little behind the mouth; they do not
appear to be used in giving motion, which is effected
by the tail, but in balancing the body, and supporting the young.

The general colour of the animal is very dark
grey, nearly black on the upper parts, but more silvery beneath. Old males usually have a large spot
of pale grey on the front of the head, when they are
said to be grey-headed. The motions of these enormous creatures are exceedingly curious: when moving perfectly at leisure, the Whale swims slowly
along, just below the surface of the water, effecting

his progress by gently moving his tail from side to side obliquely. The bunch and hump may be seen above the water, and by the disturbance which they cause in cutting the fluid, some foam is produced, by which an experienced whaler can judge, even at some miles' distance, how fast the animal is going. When disturbed, however, or from any cause inclined to increase his velocity, he uses a very different mode of progression. The broad tail now strikes the water upward and downward alternately with great force; at every blow downward the fore part sinks down several yards into the water, while by the force of the upward blow the head is thrust entirely out of the water. A Whale can swim in this manner, the head alternately appearing and disappearing, which the seamen call "going head-out," at the rate of twelve miles an hour. It may appear surprising that so bulky a portion of the animal as the enormous head, should be so easily thrust into the air, the head being usually the heaviest part of an animal: but here we trace the beneficent hand of God in creation, the volume of the head being occupied not with dense bone, but, as we have seen, with an oil which is considerably lighter than water, and which renders this part the most buoyant of the whole body. And when we consider that the breathing aperture, or blow-hole, must be projected from the water for the reception of air, we see the reason of this buoyancy.*

* For most of the particulars of the history and pursuit of this animal I am indebted to Mr. Beale's valuable work on the Sperm Whale

Every thing connected with the breathing of the Sperm Whale is performed with a regularity that is very remarkable. The length of time he remains at the surface, the number of "spoutings" made at each time, the length of interval between the spouts, the time he remains below the surface, before again rising to breathe, are all, when he is undisturbed, as regular in succession and duration as it is possible to imagine. This is a circumstance of the greatest value to the whaler; for though there is considerable variation in these particulars in different animals, yet such is the precision with which each maintains his own rates of movement, that when the periods of any particular Whale have been observed, the whaler can calculate, even to a minute, when it will reappear, and how long it will continue at the surface. A large male, called "a bull whale," usually remains at the surface about ten minutes, during which he spouts sixty or seventy times; then, to use the nautical phrase, "his spoutings are out," the head gradually sinks, the "small" is projected from the water, and presently the "flukes" of the tail are raised high in the air, and the animal descends perpendicularly to an unknown depth, remaining below from an hour to an hour and twenty minutes, when he comes up to respire again.

The regular recurrence of these motions can be depended on only when the Whale is perfectly at ease; for, if alarmed, he dives immediately, rising, however, soon again to complete his spoutings. When "going head out," also, he spouts at every projection of the head, and much more hurriedly than usual. One would be apt to suppose that a creature so huge and powerful, would be little the subject of fear or alarm; but, in truth, it is a remarkably timid animal; the approach even of a boat causing him to descend with precipitation. It is graciously ordained, that the creatures which are formed to contribute to man's comfort or sustenance, though many of them are more powerful than he, should be impressed with such a fear of him, as in general to be incapable of using their superior strength to his disadvantage. "And the fear of you, and the dread of you, shall be upon every beast of the earth, and upon every fowl of the air; upon all that moveth upon the earth, and upon all the fishes of the sea; into your hand are they delivered."* But this huge animal has other enemies than man: equally with the Greenland Whale, it is subject to the assaults of some of the larger predaceous fishes; the Sword-fish and the Saw-fish plunge into his body their formidable snouts, and the "Thresher" leaps upon him from above. Mr. Beale records the following incident, as reported to him by an eye-witness, a gentleman on whose veracity he could rely. "He stated that he had been observing a Sperm Whale during the time it had remained at the surface to breathe, which afterwards went through the evolution of peaking its flukes in the usual manner, and disappeared. As it was a large Whale, and as he knew it was likely to remain under water for a considerable time, he scarcely expected to see it again. However, in this he was mistaken;

* Gen. ix. 2.

for after it had disappeared only for a few minutes,
it again rose, apparently in great trepidation, and,
as it reared with great velocity, half of its huge
body projected out of the water. Gaining, however,
in a few seconds the horizontal position, it went on
at its utmost speed, going head out; the moment
after which, he saw a fish, somewhat resembling
a Conger-eel in figure, but rather more bulky, and
to all appearance about six or eight feet in length,
flying itself high out of the water after the Whale,
and fall clumsily on its back, which caused still
more alarm to the immense but timid animal, so
that it beat the water with its tail, and reared its
enormous head so violently, that sounds from the
former could be heard at a great distance: it still,
however, continued its rapid career, receiving every
few minutes the unwelcome visits of its galling
adversary. My informant had good reason to believe that some other animal was at the same time
attacking it from below; for, on more than one
occasion, he saw some animal dart at times to the
surface with amazing quickness, as if engaged with
great fury in the contest; and which, he supposes,
prevented the Whale from descending, in which he
had the power, no doubt, if he had not been thus
prevented, of leaving his antagonists far behind.
The attack was continued for a considerable time,
during which the Whale had got a great distance
from the ship, when it twice threw itself completely
out of its native element, no doubt endeavouring
to escape from its tormenting adversaries by this
act of 'breaching,' and which I have myself seen

him do, after having been unsuccessfully chased by the boats."*

A Whale will occasionally place himself perpendicularly in the water, his whole head being visible, presenting a most extraordinary appearance, like a black rock in the Ocean: the object of this posture is to take a rapid and comprehensive glance around him, when he is apprehensive of danger. Sometimes, when attacked by boats, he will carefully sweep his tail from side to side upon the surface, as if to discover by feeling, the object of his dread. At other times, he amuses himself by lashing the water with the same organ, in the most violent manner; covering the sea with foam, while the strokes resound on every side. Breaching or leaping bodily into the air, is alluded to in the above extract.

The food of the Sperm Whale consists of different species of cuttle or squid, occasionally varied with small fish: to obtain these, Mr. Beale supposes with much probability, that he descends to a considerable depth, and remaining as quiet as possible, allows his narrow lower jaw to hang down perpendicularly at right angles with his body. The whole inside of his mouth, and particularly the teeth, being of a glistening white hue, the squid are attracted to visit it, and when a sufficient number are within, the mouth is supposed to be closed. That the prey is obtained in some other way than by pursuit is proved by the fact, that Whales are often found blind, and others with the lower jaw distorted, which yet are in as good condition as others. These dis-

* Beale's Sperm Whale, p. 49.

tortions arise from battles between old "bull whales;'
they rush upon each other with great fury, their
mouths wide open, each endeavouring to seize his
adversary by the lower jaw. In this manner they
often become locked together by the jaws, and then
struggling with all their gigantic power, the contest
frequently terminates in the dislocation or fracture
of the jaw. The teeth are not used for chewing, the
prey being swallowed entire.

In the chase and capture of this immense creature,
as might be expected from the peculiarities of its
habits, there are several circumstances that distinguish it from the Greenland whale-fishery, while,
at the same time, there is a general resemblance.
Ships of three or four hundred tons are selected
for the voyage, strongly built, manned with a crew
of about thirty hands, and provisioned for four years.
A watch is stationed aloft immediately on leaving the
Channel, although the Sperm Whale is rarely seen in
the Atlantic north of the equator. The look-out on
the mast-head is never interrupted during the voyage,
or until the cargo is completed, the men on this duty
being relieved in succession. On a Sperm Whale
being perceived, the intelligence is communicated by
the watch calling out aloud in a peculiar tone, "There
she spouts!" a cry which fails not to produce a general rush on deck of all hands. The captain eagerly
asks, "Whereaway?" The position of the prey is
pointed out, while at every fresh spouting the watch,
accompanied by every individual on board who has
caught sight of the object, vociferates, "There again!"
When the spoutings are out, and the Whale descends,

the elevation of the tail into the air is announced
in the same manner by "There goes flukes!" The
reason of these announcements appears to be, that
the times of the animal's motions may be accurately
marked by the proper officers, though they may not
see them themselves, as affording an unfailing cri-
terion by which to judge of his future movements.
On the first signal being given, the boats, which are
always kept in complete readiness at the ship's side,
are lowered, and the men take their places with joyous alacrity. If not too far off, they strain every
nerve to arrive at the animal before his spoutings
are out, which in a large bull Whale may be about
ten minutes. Should they be unable, however, to
effect this, they endeavour to mark his direction of
diving, and station themselves near the spot where
they expect he will break water. On his reappear-
ance, the boats are rowed up as silently as possible,
and the foremost harpooner darts his weapon with
all his force into his side. The instant this is done
he cries, "Stern all!" and the boat is withdrawn with
precipitation. The Whale, writhing with the agony,
dives perpendicularly, drawing the line of the harpoon swiftly through its groove: the other boats
are ready to bend on their lines, each of which is
two hundred fathoms long; for sometimes a Whale
will drag after him four lines descending to the
depth of 4800 feet. Presently he is seen approaching the surface: "The gurgling and bubbling water,
which rises before, also proclaims that he is near;
his nose starts from the sea; the rushing spout is
projected high and suddenly, from his agitation."

On his reaching the surface, the other boats infix their harpoons, while at the same instant the former harpooner thrusts deeply his steel lance into the body, and "Stern all!" again resounds.

Now comes the most dangerous part of the business; the Whale is in his "flurry," or last agony; he dashes hither and thither, snaps convulsively with his huge jaws, rolls over and over, coiling the line around his body, or leaps completely out of the water. The boats are often upset, sometimes broken into fragments, and the men wounded or drowned. Now the crimson blood is spouted from the blow-hole, and falls in showers around; the poor animal whirls rapidly round in unconsciousness, in a portion of a circle, rolls over on its side, and is still in death.

The huge body is now towed to the ship; a hole is cut into the blubber near the head, into which a strong hook is inserted; a difficult and dangerous operation. A strong tension is then applied to this hook, and by it the blubber is hoisted up, as it is gradually cut by the spades in a spiral strip, going round and round the body. As this strip or band of blubber is pulled off, the body of course revolves, until the stripping reaches "the small," when it will turn no more. The head, which at the commencement of the process was cut off and secured astern, is now hoisted into a perpendicular position, the front of the muzzle opened, and the spermaceti dipped out of the "case" by a bucket at the end of a pole. The "junk" is then cut into oblong pieces, and the remainder of the head, with the carcass, cut adrift.

The oil is afterwards extracted from the blubber and junk by exposing them to the action of fire in large pots, the skinny portions which remain serving for fuel: and the spermaceti is purified in the same manner. The products are then stowed away in barrels in the vessel's hold.

The following narrative, from the interesting work of Mr. Beale, gives us a vivid picture of this exciting pursuit: "At daybreak, one fine morning in August, as our first mate was going aloft to look out for Whales, he discovered no less than three ships within a mile of us; but they were situated in various directions. We soon discovered them to be whalers, who, like ourselves, were cruising after the Spermaceti Whale, and, therefore, their appearance only had the effect of redoubling our vigilance in the look-out, so that we might, if possible, be the first to obtain the best chance, if one of those creatures hove in sight. And it was not long before a very large Whale made his appearance right in among the ships. The water was smooth at the time, for we had but a light air of wind stirring, so that our boats were instantly lowered without the loss of time of bringing the ship to. But, although we managed matters as quietly and secretly as possible, we found the moment our boats quitted the ship's side, that all the others had been as vigilant as ourselves, and had also lowered their boats after the Whale. The whole of them immediately began the chase, nine boats in all, being three from each ship. They all exerted themselves to the utmost, and, as we expected, in vain; for before any

Y

of the boats had got even near him, the enormous
animal lifted his widely-expanded flukes, and descended perpendicularly into the depths of the Ocean
to feed. Those in the boats, however, having noticed his course, proceeded onwards, thinking the
Whale would continue to pursue the same direction
under water; but, as he was going slowly at the
time he was up, they did not proceed more than a
mile from the place at which he descended, before
they separated about a hundred yards from each
other, and then, peaking their oars, all the men in
each boat stood up, looking in different directions,
so as to catch the first appearance of the spout, when
the Whale again rose to breathe. When an hour
after his descent had expired, the excitement among
us who were on board the ship, became wound up
to its highest pitch. The captain, who had remained
on board, ascended to the fore-top-gallant-yard to
watch the manœuvres of the boats, and for the
purpose of the better ordering the signals to them,
or working of the ship. All those who were down
after the Whale appeared as feverish with anxiety
as ourselves, for every now and then they were to
be seen shifting their position a little, thinking
to do so with advantage; then they would cease
rowing, and stand up on the seats of the boats, and
look all round over the smooth surface of the Ocean
with ardent gaze. But one hour and ten minutes
expired before the monster of the deep thought
proper to break cover; and when he did, then a
rattling chase commenced with the whole of the
boats, and they really flew along in fine style, some

of them appearing to be actually lifted quite on the surface of the water, from the great power of the rowers; and we had the satisfaction of observing, that our boats were quite equal to the others in the speed with which they were propelled. But it was again a useless task, as the Whale had outwitted those in the boats, by having gone, while under water, much further than any of his pursuers had anticipated, and they again had the mortification of witnessing the turning of his flukes, as he once more descended into the depths of his vast domain. We now knew to a minute the time that he would remain below, while the people in the boats continued to row slowly onwards the whole time. A fine breeze now sprang up, so that we were enabled to keep company with the boats, keeping a little to windward of them, as the Whale was going 'on a wind,' as a seaman would say, meaning that it was blowing across him.

"When the hour and ten minutes had again nearly past, the nine boats were nearly abreast of each other, and not much separated, so that the success of first striking the Whale depended very much upon the swiftest boat, especially if the Whale came up ahead. We had now all the boats on our lee-beam, while the ships were all astern of us, the most distant not being more than half a mile, so that we enjoyed an excellent view of this most exciting and animated scene. True to his time, the leviathan at length arose right ahead of the boats, and at not more than a quarter of a mile distant from them. The excitement among the crews of the various

boats, when they saw his first spout, was tremen·
dous; they did not shout, but we could hear an agitated murmur from their united voices reverberating
along the surface of the deep.   They flew over the
limpid waves at a rapid rate: the mates of the various boats cheered their respective crews by various
urgent exclamations.   'Swing on your oars, my
boys, for the honour of ·the Henrietta!' cried one;
'Spring away, hearties!' shouted another; and yet
scarcely able to breathe from anxiety and exertion;
'It's our fish!' vociferated a third, as he passed the
rest of his opponents but a trifling distance.   'Lay
on, my boys!' cried young Clark, our first mate, as
he steered the boat with one hand and pressed down
the after oar with the other: 'she'll be ours yet;
let's have a strong pull, a long pull, and a pull
all together!' he exclaimed, as he paused from his
exertions at the after oar, which soon brought up
his boat quite abreast of the foremost.

"But the giant of the Ocean, who was only a
short distance before them, now appeared rather
'gallied,' or frightened, having probably seen or
heard the boats, and as he puffed up his spout to
a great height, and reared his enormous head, he
increased his speed, and went along quite as fast as
the boats, but for only two or three minutes, when
he appeared to get perfectly quiet again, while the
boats gained rapidly upon him, and were soon close
in his wake. 'Stand up!' cried young Clark to
the harpooner, who is also the bow-oarsman; while
the same order was instantly given by his opponent,
whose boat was abreast of our mate's with the rest

close to their sterns. The orders were instantly obeyed, for in a second of time both boat-steerers stood in the bows of their respective boats, with their harpoons held above their heads ready for the dart; but they both panted to be a few yards nearer to the Whale, to do so with success. The monster plunged through the main quickly, but the boats gained upon him every moment, when the agitation of all parties became intense, and a general cry of 'Dart! dart!' broke from the hindermost boats, who each urged their friends, fearful of delay. The uproar became excessive, and while the tumult of voices, and the working and splashing of the oars, rolled along the surface of the deep, both the harpooners darted their weapons together, which, if they had both struck the Whale, would have originated a contention between them, regarding their claims. But, as it happened, neither of them had that good fortune; for, at the moment of their darting the harpoons, the Whale descended like a shot, and avoided their infliction, leaving nothing but a white and green-looking vortex in the disturbed blue Ocean, to mark the spot where his monstrous form so lately floated. A general huzza burst from the sternmost boats, when they saw the issue of this chase, thinking, now, that another chance awaited them on the next rising of the Whale, and they soon began to separate themselves a little, and to row onwards again in the course which they thought he had taken. Our captain, feeling irritated at the ill-success of the mate, now ordered his own boat to be lowered, intending to make one in the chase him-

self; but, just as he had parted from the ship, going
down a little to leeward, a tremendous shout arose
from the people in our own boats, joined with a loud
murmuring from the rest of the boats' crews; for
the Whale, not having had all its spoutings out, had
now risen again to finish them, and was coming to
windward at a quick rate, right towards our ship.
The captain saw his favourable situation in a moment, and passing quickly to the bows of the boat,
he stood to waylay him as he came careering along,
throwing his enormous head completely out of the
water, for he was now quite 'gallied.' He soon
came, and caught a sight of the boat just as he
got within dart; the vast animal rolled himself
over in an agony of fear, to alter his course; but
it was too late; the harpoon was hurled with excellent aim, and was plunged deeply into his side,
near the fin.

"As the immense creature almost flew out of the
water from the blow, throwing tons of spray high
into the air, showing that he was 'fast,' a triumph-
ant cheering arose from those in our own boats, as
well as from those in the ship, accompanied by exclamations loud and deep, and not of the most favourable kind to us, from all the rest. But onwards
they all came, and soon cheerfully rendered assistance to complete its destruction; but which was not
done, however, without considerable difficulty, the
Whale continuing to descend the moment either of
the boats got nearly within dart of him. But after
an hour's exertion in this way, six out of the ten
boats which were now engaged got fast to him by

their harpoons, but not one of them could get near enough to give him a fatal lance. He towed them all in various directions for some time, taking care to descend below the surface the moment a boat drew up over his flukes, or otherwise drew near, which rendered it almost impossible to strike him in the body, even when the lance was darted, although the after part of his 'small' was perforated in a hundred places: from these wounds the blood gushed in considerable quantities, and as the poor animal moved along, towing the boats, he left a long ensanguined stain in the Ocean. At last, becoming weak from his numerous and deep wounds, he became less capable of avoiding his foes, which gave an opportunity for one of them to pierce him to the life! Dreadful was, that moment, the acute pain which the leviathan experienced, and which roused the dormant energies of his gigantic frame. As the life-blood gurgled thick through the nostril, the immense creature went into his 'flurry' with excessive fury; the boats were speedily sterned off, while he beat the water in his dying convulsions with a force that appeared to shake the firm foundation of the Ocean."*

Few occurrences in a long voyage are more generally interesting and exciting than the sight, and particularly the speaking, of another ship. Even in crossing the Atlantic this is the case; but how much more in a voyage to the Pacific, where many months may elapse without the appearance of a vessel! Tho

* Hist. of Sperm Whale, p. 176.

call of "Sail ho!" has an electric effect: all the
telescopes on board are soon pointed towards her;
her rig, her canvas, her direction, the force of wind
she has, the tack she is on, if "by the wind," are
all carefully scrutinized and commented on. If the
courses of the two vessels, and their positions, are
such that they will approach very near to each other,
they will "speak," as a matter of course; but there
are few commanders so churlish as not to submit
to a slight deviation of their course in order to com-
municate with another. Perhaps the stranger is
seen directly astern, following right in the wake, a
circumstance which, as far as my own observation
extends, commonly excites a slight feeling of un-
easiness, and a more than usual attention to her appearance, powers of sailing, &c. Though the reason
assures one that the occurrence of a ship in that
particular direction, is as likely as in any other
quarter, yet the mind will recur to the idea of pursuit, and thoughts of walking the plank, or hanging
at the yard-arm, will crowd up to the imagination,
especially if the locality happen to be the West
Indies, or the Spanish Main, or any other sea habitually infested with pirates. But as she gains
a greater nearness, her hull and rig indicate her to
be a peaceful trader, and presently the bunting is
run up to the peak, and the folds of England's fair
ensign flow out upon the breeze. The approach
of a vessel is always a pleasing sight; her graceful
movements, as she bounds over the waves, the white
foam rolling up under her bows, her taper masts
and spars, the elegant curves which the breeze gives

to her running rigging, the white, plump sails, belly-
ing from the wind, are all beautiful; if she is to
windward, her clean white decks are visible as she
lies over, the crew collected in the waist, or about
the bows, the officers and passengers assembled on
the quarter-deck, gazing with equal curiosity to our
own, upon our appearance; the captain standing
with his speaking-trumpet in his hand ready to seize
the moment of nearest approach. He raises his
trumpet to his mouth—"Ship ahoy!" "Hilloa!"
"What ship is that, pray? Where are you from?
Where are you bound? How long are you out?
What's your longitude?" These and similar questions are mutually asked and answered, each reply
being acknowledged by a slight motion of the trumpet in the air. If there be opportunity, the pre-
vailing character of the winds with each, the prospects of the voyage, the state of the respective
crews, and other nautical subjects, are interchanged;
but usually the time afforded for speaking by the
vessels remaining within hail, is very brief, and they
again diverge, and soon are lost to each other below
the horizon. Very often, from the sighing of the
wind among the cordage, the working of the ship,
the ripple and splash at her side, as well as from
distance, while the questions from being so much
in course, are perfectly intelligible, the answers are
almost inaudible, and can sometimes only be guessed
at, the consonants being entirely lost, and the vowel-
sounds alone heard. This will explain a laughable
incident which took place a few years ago, on the

homeward passage of the John Bull transport, from Rio Janeiro.

One fine starlight evening, about half-past eight o'clock, the officer on deck came into the cabin, and announced that a ship was hailing. All hands immediately came on deck, and the captain asked the position of the stranger. At that moment, "Ship ahoy!" was heard, the voice apparently being to windward. A lantern was put over the gangway, the mainsail was hauled up, and the mainyard backed, to stop the vessel's way. No ship was to be seen. "Silence, fore and aft!" ordered the captain, for the decks were now crowded, soldiers, sailors, women, children, all were up. "Ship ahoy!" again came over the waves, and "Hilloa!" answers the captain at the top of his voice. Every one now listened with breathless attention for the next question, expecting the name of the ship would be demanded, as usual: "Ship ahoy!" again resounded, and several together answered "Hilloa!" louder than before: but no notice was taken of the reply, and no sail was in sight. "It is very strange!" exclaimed the captain; "where can she be?" One thought she might have passed them; others suggested that it might be a pirate-boat about to board. The captain took the hint, put the troops under arms, cleared away the guns ready for action, and double-shotted them. Silence being again obtained, "Ship ahoy!" was heard again, and the voice still seemed to come from the windward. The chief mate then suggested the possibility of some person being on a raft, and volunteered to go in a boat to ascertain. The boat was lowered, and the two mates, with the boat's crew, each armed with sword and pistol, rowed at some distance round the ship.

On the officer's return, they reported that they could neither hear nor see any thing. Silence prevailed while they reported this to the captain, every one being desirous to know the issue of the search. Instantly, the same "Ship ahoy!" was heard, though much less audibly, and, apparently, at a greater distance than before. The next moment it was heard much louder and closer. A feeling of intense excitement now prevailed in each of the crowd of persons on board the transport. More than an hour had passed since the ship was hove to; every one had repeatedly heard the stranger's hail, coming through the darkness, but nothing had been seen of him, and no further question or answer could be elicited. The screams of the women and children, and the muttering of the men, showed that superstitious dread of something supernatural and unearthly was creeping over every one. The captain issued orders to shoulder arms and to make ready the guns.

Just at this crisis, one of the cabin-boys, who had been standing near the mainmast, stepped aft to the chief mate, and said, "It's a fowl in the hencoop, sir, that's a-making that 'ere noise." That officer indignantly bestowed on him a sound box on the ear for his information, but immediately recollecting that he was an intelligent lad, accompanied him to the hencoop with a lantern; where he saw a fowl lying on its side. He took it out, and placed it on the capstan; and there, in the sight of the whole company, was beheld a poor hen dying of the croup, occasionally emitting a sound "ee-a-aw," which resembled the words "Ship ahoy!" coming from a distance, as closely as any hail that was ever heard.*

* Naut. Mag. 1842, p. 409.

THE PACIFIC OCEAN.

A REMARKABLE feature in the Pacific Ocean, and one that distinguishes it from every other sea, is the immense assemblage of small islands with which it is crowded, particularly in the portion situated between the tropics. For about three thousand miles from the coast of South America, the sea is almost entirely free from islands; but thence to the great isles of India, an immense belt of Ocean, nearly five thousand miles in length, and fifteen hundred in breadth, is so studded with them as almost to be one continuous archipelago. The term Polynesia, by which this division of the globe is now distinguished, is compounded of two Greek words, signifying *many islands*. Very few of these gems of the Ocean are more than a few miles in extent, though Tahiti, and some in the more western groups, are of rather larger dimensions; while Hawaii, the largest island in Polynesia, is about the size of Yorkshire.

The isles, which in such vast numbers thus stud the bosom of the Pacific, are of three distinct forms, the Coral, the Crystal, and the Volcanic. Of these, the first formation greatly predominates; but the largest islands are of the last description: of the crystal formation but few specimens are known.

Z   (265)

Imagine a belt of land in the wide Ocean, not more
than half a mile in breadth, but extending, in an
irregular curve, to the length of ten or twenty miles
or more: the height above the water not more than
a yard or two at most, but clothed with a mass of
the richest and most verdant vegetation. Here and
there, above the general bed of luxuriant foliage,
rises a grove of cocoa-nut trees, waving their feathery plumes high in the air, and gracefully bending
their tall and slender stems to the breathing of the
pleasant trade-wind. The grove is bordered by a

CORAL ISLAND.

narrow beach on each side, of the most glittering
whiteness, contrasting with the beautiful azure

waters by which it is environed. From end to end of the curved isles stretches, in a straight line, forming, as it were, the cord of the bow, a narrow beach, of the same snowy whiteness, almost level with the sea at the lowest tide, enclosing a semi-circular space of water between it and the island, called the lagoon. Over this line of beach, which occupies the leeward side, the curve being to windward, the sea is breaking with sublime majesty; the long unbroken swell of the Ocean, hitherto unbridled through a course of thousands of miles, is met by this rampart, when the huge billows, rearing themselves upwards many yards above its level, and bending their foaming crests, "form a graceful liquid arch, glittering in the rays of a tropical sun, as if studded with brilliants. But, before the eyes of the spectator can follow the splendid aqueous gallery which they appear to have reared, with loud and hollow roar they fall, in magnificent desolation, and spread the gigantic fabric in froth and spray upon the horizontal and gently broken surface." Contrasting strongly with the tumult and confusion of the hoary billows without, the water within the lagoon exhibits the serene placidity of a mill-pond. Extending downwards to a depth, varying from a few feet to fifty fathoms, the waters possess the lively green hue common to soundings on a white or yellow ground; while the surface, unruffled by a wave, reflects with accurate distinctness the mast of the canoe that sleeps upon its bosom, and the tufts of the cocoa-nut plumes that rise from the beach above it. Such is a Coral Island, and if its appearance is one of singular loveli-

ness, as all who have seen it testify, its structure,
on examination, is found to be no less interesting
and wonderful. The beach of white sand, which
opposes the whole force of the Ocean, is found to
be the summit of a rock which rises abruptly from
an unknown depth, like a perpendicular wall. The
whole of this rampart, as far as our senses can
take cognizance of it, is composed of living coral,
and the same substance forms the foundation of the
curved and more elevated side which is smiling in
the luxuriance and beauty of tropical vegetation.
The elevation of the coral to the surface is not
always abruptly perpendicular; sometimes reefs of
varying depths extend to a considerable distance
in the form of successive platforms or terraces. In
these regions may be seen islands in every stage
of their formation: " some presenting little more
than a point or summit of a branching coralline
pyramid, at a depth scarcely discernible through the
transparent waters; others spreading, like submarine
gardens or shrubberies, beneath the surface; or
presenting here and there a little bank of broken
coral and sand, over which the rolling wave occasionally breaks;" while others exist in the more
advanced state that I have just described, the main
bank sufficiently elevated to be permanently protected from the waves, and already clothed with
verdure, and the lagoon enclosed by the narrow
bulwark of the coral reef. Though the rampart thus
reared is sufficient to preserve the inner waters in
a peaceful and mirror-like calmness, it must not
be supposed that all access to them from the sea

is excluded. It almost invariably happens that, in the line of reef, one or more openings occur, which, though sometimes narrow and intricate, so as scarcely to allow the passage of a native canoe, are not unfrequently of sufficient width and depth to permit the free ingress of large ships. This is a very remarkable instance of the Divine care over the little creatures which rear these solid structures; they appear to be endowed with an instinctive knowledge, that if the reef were carried uninterruptedly along from one point to another, so as completely to shut in the lagoon, the water within would soon become unfit to support their existence, and would ultimately be dried up. The advantage to man of these openings is very great; without them the islands might smile invitingly, but in vain; no access could be obtained to them by shipping, through the tremendous surf by which their shores are lashed; but by these entrances the lovely lagoons are converted into the most quiet, safe, and commodious havens imaginable, where ships may lie, and wood and water, and refresh their crews, in security, though the tempest howl without. It is a scarcely less beneficent provision that the position of the openings is in most cases indicated so as to be visible at a great distance. Had there been merely an opening in the coral rock, it could not have been detected from the sea, except by the diminution of the foaming surf just at that spot; a circumstance that could scarcely be visible, unless the observer were opposite the aperture. But, in general, there is on each side of the passage, a little islet, raised

z 2

on the points of the reef, which, being commonly
tufted with cocoa-nut trees, is perceptible as far off
as the island itself, and forms a most convenient
landmark.

Notwithstanding that the highest point of these
narrow islets is rarely more than a yard above the
tide, it is a remarkable fact that fresh water is frequently found in them. It is probable that the coral
rock acts as a filter, allowing the sea-water to perco-
late through its porous substance, but excluding all
its saline particles held in solution.

Though I have described the two parts of a Coral
Island, or Atoll, as it is called, as distinct, yet the
difference is only in appearance; the foundation on
every side is the same, a coral reef rising to the surface : but the side most exposed to the action of the
waves driven in by the trade-winds, is invariably the
first to be projected, and attains a higher elevation
than the leeward side. Neither must it be supposed
that the belt to windward is always continuous,
though the interruptions are comparatively few. A
close inspection will likewise show that the outline
of the whole reef possesses much less regularity of
form than its aspect from a distance indicated. The
form, however, is invariably a more or less close
approach to a circle. Sometimes the land is continuous through the whole circumference, with the
exception of a channel or two into the lagoon, which
presents the appearance of a circular pond with a
verdant border surrounding it; again, another atoll
will be found which has brought its ring of reef
scarcely to the surface, exposing, perhaps, a single

bare spot on the windward edge at the lowest ebb of spring tide.

Captain Basil Hall has recorded some pleasing observations on this singular formation, in his voyage to Loo-Choo. He says—

"The examination of a coral reef during the different stages of one tide, is particularly interesting. When the sea has left it for some time, it becomes dry, and appears to be a compact rock, exceedingly hard and rugged; but no sooner does the tide rise again, and the waves begin to wash over it, than millions of coral worms protrude themselves from holes on the surface, which were before quite invisible. These animals are of a great variety of shapes and sizes, and in such prodigious numbers that in a short time the whole surface of the rock appears to be alive and in motion. The most common of the worms at Loo-Choo was in the form of a star, with arms from four to six inches long, which it moved about with a rapid motion in all directions, probably in search of food. Others were so sluggish that they were often mistaken for pieces of the rock; these were generally of a dark colour, and from four to five inches long, and two or three round. When the rock was broken from a spot near the level of high-water, it was found to be a hard, solid stone; but if any part of it were detached at a level to which the tide reached every day, it was discovered to be full of worms, all of different lengths and colours, some being as fine as a thread, and several feet long, generally of a very bright yellow, and sometimes of a blue colour; while others resembled

snails, and some were not unlike lobsters or prawns
in shape, but soft, and not above two inches long."*
Some of the animals thus described by the Captain,
were doubtless intruders that had sought shelter or
food in the interstices of the coral: the true archi-
tects of these wonderful structures are polypes of
minute size, which, though of many varying species,
and even genera, agree in the simplicity of their form
and structure. They consist of a little oblong bag
of jelly, closed at one end, but having the other
extremity open, and surrounded by tentacles, usually
six or eight in number, set like the rays of a star.
Multitudes of these tiny creatures are associated in
the secretion of a common stony skeleton, the coral,
or madrepore; in the minute orifices of which they
reside, protruding their mouths and tentacles when
under water, but withdrawing themselves by sudden
contraction into their holes the moment they are
molested.

It was for a long time supposed that all the islands
of coral formation were reared from their bases,
fathomless depths in the Ocean, by the unaided efforts
of these minute creatures; and from exaggerated
notions of the rapidity with which the process was
going on, anticipations were frequently uttered that
a large portion of the Pacific might, at no very distant period, be occupied by the spreading structures
united into a vast coral continent. More accurate
observations have, however, satisfactorily proved that
the living animals cannot exist at a greater depth
than twenty or thirty fathoms, so that the whole of

* Voyage to Loo-Choo, p. 75.   (Constable's edit.)

these animal secretions must have been deposited within that distance from the surface. At the same time, it is no less true that the water in the immediate vicinity of the islands is fathomless, and that the descent of their outer edge is remarkably abrupt and precipitous. The only satisfactory explanation of the phenomenon appears to be the one proposed and ably supported by Mr. Darwin, in his elaborate treatise on Coral reefs. Many islands of the common rock formation are found in the Pacific, on the shelving sides of which, a few fathoms below water, the coral animals have fixed their stony habitations, forming what is called a fringing reef, distinguished from others by being immediately attached to the land, without the intervention of any lagoon or channel of water. Mr. Darwin supposes that every island in the Pacific originally presented this structure, but that wherever a variation at present exists, the solid rock has been gradually, and perhaps very slowly, subsiding to a lower level. Now, let us assume this state of things for a moment, and look at the results. We must, however, mention two well-ascertained instincts of the Polype: the one is, that it works up towards the light; the other, that its proceedings are most vigorous at the outer edge, where it is washed by the beating waves. Let A represent the section of a rocky island; B, B, the level of low-water; and D, the reef of coral fringing the coast. After the lapse of time, during which it has been subsiding, the water-level stands at *b, b;* the coral at D has died from the too great depth, but the animals have been working upwards upon the dead

18

matter, so that living coral is still near the surface;
the superior vigour of the species inhabiting the seaward edge, however, has caused that edge to be more

SECTION OF CORAL ISLAND.

elevated than the interior, as at d, d; so that the
appearance is now that of a rocky isle, diminished in
extent, surrounded by a reef at some distance, separated by the intervention of a shallow channel, e, e:
this is exactly the appearance of Tahiti and the
larger islands generally, as I shall mention more fully
when I come to the volcanic formation. The subsidence still goes on; and, after a while, the water,
β, β, is level with the summit of the island, which, of
course, is now an island no longer; the growth of the
coral has kept pace with the depression, and it is
still at the surface, as at δ, δ; the more slowly growing species of the interior are still overflowed, and, as
the island is submerged in the centre, the water, ϵ, ϵ,
is no longer an annular channel, but a round lagoon;
and thus we have an atoll, as at first described. The
subsequent process of elevating and clothing the new
islets is a rapid one. Chamisso observes, " As soon
as it has reached such a height that it remains
almost dry at low-water at the time of ebb, the

corals leave off building higher; sea-shells, fragments of coral, sea-hedgehog shells, and their broken-
off prickles, are united by the burning sun through
the medium of the cementing calcareous sand, which
has arisen from the pulverization of the above-men-
tioned shells, into one whole or solid stone, which,
strengthened by the continual throwing up of new
materials, gradually increases in thickness, till it at
last becomes so high that it is covered only during
some seasons of the year by the spring-tides. The
heat of the sun so penetrates the mass of stone when
it is dry, that it splits in many places, and breaks off
in flakes. These flakes, so separated, are raised one
upon another by the waves, at the time of high-
water. The always-active surf throws blocks of coral
(frequently of a fathom in length, and three or four
feet thick), and shells of marine animals, between
and upon the foundation stones. After this the calcareous sand lies undisturbed, and offers to the seeds
of trees and plants cast upon it by the waves, a soil
upon which they rapidly grow, to overshadow its
dazzling white surface. Entire trunks of trees,
which are carried by the rivers from other countries
and islands, find here, at length, a resting-place, after
their long wanderings; with these come some small
animals, such as lizards and insects, as the first inhabitants. Even before the trees form a wood, the real
sea-birds nestle there; strayed land-birds take refuge
in the bushes; and at a much later period, when
the work has been long since completed, man also
appears, builds his hut on the fruitful soil formed
by the corruption of the leaves of the trees, and

calls himself lord and proprietor of this new creation."*

The species of Polypes which contribute to the
formation of coral structures are very numerous,
and differ greatly from each other in the forms of
their respective habitations. Some form large rounded masses, with numerous winding depressions, as
the Brainstones (*Meandrina*); some are studded with
holes, filled with thin shelly plates placed perpendicularly, and converging to a point in the centre,
as *Astræa;* some assume the appearance of a mush-
room, as *Agaricia;* but the most general form is
that of an irregular, branching shrub. The various
kinds are not found scattered indiscriminately over
the whole edifice, but each occupying its own zone
and position, each performing its own part, assigned
by God, in carrying up the wondrous architecture.
The principal and most important place is filled by
the genus *Porites*, which occupies the outside of the
reef, at the exposed edge, constructing large rounded
masses. The next in importance is the *Millepora
complanata*, which forms thick vertical plates, uniting at different angles by their edges, so as to present the appearance of a honeycomb: the marginal
plates only being alive. These two kinds alone
are able to endure the intermitting exposure to
which the upper edge is subject, in being continually washed over by the surf; other species are
found a few fathoms down. Inside the lagoon,
there are quite distinct sorts, generally brittle, and
thinly branched; while great round Brainstones

* Kotzebue's Voyage.

(*Meandrina*), and flower-like *Caryophilla*, occupy the bottom. In the shallow hollows of the reef, *Pocillopora verrucosa*, a species having short waved plates or branches, is found: when alive it is a beautiful object, being of a delicate pale crimson hue.

Conflicting statements have been made respecting the activity of the building processes going on in the present age; some affirming that the reefs have acquired no perceptible addition, either to their height or extent, since they have been known; others anticipating a speedy filling up of the Pacific from their rapid growth. The truth seems to be, that, while in some localities no change in extent can be traced through many years, in others very rapid enlargements are made. As showing the rate at which coral grows under favourable circumstances, Mr. Darwin mentions two or three interesting cases. In the lagoon of Keeling Atoll, a channel was dug, for the passage of a schooner built upon the island, through the reef into the sea; in ten years afterwards, when it was examined, it was found almost choked up with living coral. Dr. Allan, at Madagascar, placed several masses of coral, of different species, each weighing ten pounds, in the sea three feet beneath the surface, where they were secured from removal by stakes. This was in December; and in the month of July following, they were found nearly extending to the surface, immovably fixed to the rock, and grown to several feet in length. A ship in the Persian Gulf, in the course of twenty months, had her copper encased with living coral to the thickness of two feet.

2 A

It may excite surprise, that the openings in the
reefs are not gradually filled up in those cases
where no stream of fresh water flows into the sea.
But it appears that the presence of any sediment
is so annoying to the animals, as to prevent their
acting with energy. This may be produced in
various modes: there are many animals which
feed on the living coral. Mr. Darwin observed
two Parrot-fishes (*Scarus*), one outside and the
other inside the reef, both engaged in devouring
it: many small Mollusca penetrate into it, and
the Sea-cucumbers (*Holuthuria*), which are very
numerous and large, are continually nibbling at it.
The rolling of dead masses by the surf must also
chafe away particles continually, and the presence
of the deposited sand thus formed is doubtless one
reason why the coral grows languidly within the
lagoon; whereas the abraded atoms on the outside
are at once washed off by the waves, and sink to
the bottom of the Ocean. Now, the water which
is continually thrown into the lagoon by the surf
breaking over the reef, can find an outlet only
through the openings of which I am speaking; and
thus a constant current is maintained through them,
and particularly at the sides, where the opposing
waves offer less resistance, carrying out some of the
sediment, and depositing it in its course on the
coral margins of the aperture. The coral sand made
by these abraded fragments is quickly cemented
by the influence of the sun into a solid mass, where
exposed to the air; and it is, perhaps, owing to this
property that the numberless little islets are formed

along the reef, even where there is no aperture. The surf in violent gales can roll up upon the reef masses of torn-off coral, weighing many hundred-weights; such a mass, once lodged, would be the nucleus of an islet; the sand would speedily accumulate around it, which the sun would soon cement into a mass, and then the islet would be ready for vegetation.

The following lines are beautifully descriptive of the formation of an atoll, though the author seems to hold the erroneous notion of the whole structure being elevated from the bottom by the coral polypes:—

"Millions of millions thus, from age to age,
With simplest skill, and toil unweariable,
No moment and no movement unimproved,
Laid line on line, on terrace terrace spread.
To swell the heightening, brightening, gradual mound,
By marvellous structure climbing tow'rds the day.
Each wrought alone, yet altogether wrought;
Unconscious, not unworthy, instruments,
By which a Hand invisible was rearing
A new creation in the secret deep.
Omnipotence wrought in them, with them, by them;
Hence what Omnipotence alone could do
Worms did.   *   *   *   *   *   *
    "Atom by atom thus the burthen grew,
Even like an infant in the womb, till Time
Deliver'd Ocean of that monstrous birth,
A Coral Island, stretching east and west,
In God's own language to its parent saying,
'Thus far, no farther, shalt thou go; and here
Shall thy proud waves be stayed:'—A point at first
It peer'd above those waves; a point so small,
I just perceived it, fix'd where all was floating;
And when a bubble cross'd it, the blue film

Expanded like a sky above the speck ;
That speck became a hand-breadth; day and night
It spread, accumulated, and ere long
Presented to my view a dazzling plain,
White as the moon amid the sapphire sea;
Bare at low water, and as still as death ;
But when the tide came gurgling o'er the surface,
'Twas like a resurrection of the dead;
From graves innumerable, punctures fine
In the close coral, capillary swarms
Of reptiles, horrent as Medusa's snakes,
Cover'd the bald-pate reef; then all was life,
And indefatigable industry ;
The artizans were twisting to and fro,
In idle-seeming convolutions; yet
They never vanish'd with the ebbing surge,
Till pellicle on pellicle, and layer
On layer, was added to the growing mass.
Ere long the reef o'ertopped the spring-flood's height,
And mock'd the billows when they leap'd upon it,
Unable to maintain their slippery hold,
And falling down in foam-wreaths round its verge.
Steep were the flanks, with precipices sharp,
Descending to their base in ocean-gloom,
Chasms few, and narrow, and irregular,
Form'd harbours, safe at once and perilous—
Safe for defence, but perilous to enter.
A sea-lake shone amidst the fossil isle,
Reflecting in a ring its cliffs and caverns,
With heaven itself seen like a lake below."*

The islands of the second class seem to have been
originally of the same structure as those already
noticed, but have been elevated to the height of
one hundred to five hundred feet, by some unknown
agency. The character of their vegetation resembles that of the volcanic isles, of which I shall presently speak, but they do not possess their sub-

* Montgomery's Pelican Island.

lime grandeur, nor the peculiar loveliness of the atolls. The rocks are crystallized carbonate of lime, supposed to have been originally coral, "but, by exposure to the action of the atmospheric air, together with that of the water percolating through them, the loose particles of calcareous matter have been washed away, and the whole mass has become harder and brighter." In the islands named Atiu

CRYSTAL ISLANDS.

and Mauke, the latter of which was discovered by Mr. Williams in 1823, that gentleman found several extensive caverns, having a stratum of crystallized coral, fifteen feet in thickness, as a roof. In one of these exquisitely beautiful caverns he walked about for two hours, and found no termination to its windings. This circumstance, together with

2 ▲ 2

the absence of scoria, lava, and other volcanic products, in these islands, has led him to the conclusion that they have been elevated by some expan-
sive power, or volcanic agency, without eruption.*

In one of the Tonga Isles there is a very curious
submarine cavern, connected with an interesting
legend. Mr. Mariner, who describes it, informs us
that being in the vicinity one day, a chief proposed
to visit this cave. One after another of the young
men dived into the water without rising again, and
at length the narrator followed one of them, and,
guided by the light reflected from his heels, entered a large opening in the rock, and presently
emerged in a cavern. The entrance is at least a
fathom beneath the surface of the sea at low-water,
in the side of a rock upwards of sixty feet in height;
and leads into a grotto about forty feet wide, and
of about the same height, branching off into two
chambers. As it is apparently closed on every side,
there is no light but the feeble ray transmitted
through the sea; yet this was found sufficient, after
the eye had been a few minutes accustomed to the
obscurity, to show objects with some little distinctness. Mr. Mariner, however, desirous of better
light, dived out again, procured his pistol, and after
carefully wrapping it up, as well as a torch, re-en-
tered the cavern as speedily as possible. Both the
pistol and torch, on being unwrapped, were found
perfectly dry, and by flashing the powder of the
priming, the latter was lighted, and the beautiful
grotto illuminated. The roof was hung with sta-

* Williams's Missionary Enterprises, p. 28.

lactites in fantastic forms, bearing some resemblance to the Gothic arches and carved ornaments of some old church. After having examined the curiosities' of the place, the party sat down to drink *cava*, while an old chief communicated some interesting particulars in the history of the grotto.

In former times there lived a governor of one of the neighbouring islands, who exercised his authority with the most grinding tyranny and injustice. A conspiracy against his life was formed by a subordinate chief, which was discovered, and he himself condemned to death with his family. One of his daughters, however, a beautiful girl, was reserved for a more hateful destiny, that of becoming the wife of the cruel tyrant. It happened that another young chief, who had long loved this maiden, had, a little while before, accidentally discovered the submarine cavern, when diving in pursuit of turtle. He had kept his discovery a profound secret, reserving it as a safe retreat for himself, in case he should be unsuccessful in a plan of revolt, which he also had in view. No sooner, however, were the tyrant's decisions known than he hastened to the damsel, and acquainting her with her danger, besought her to escape with him. The emergency was great; little solicitation sufficed to obtain her consent; the woods concealed her until evening, when her lover brought his canoe to a lonely part of the beach, in which she embarked with him. As he paddled her across the rippling waves, he made known to her his discovery of the grotto, in which he proposed to conceal her until they could find an opportunity for escape to a distant island. Arrived at the cliff, he conducted her through the waters to her new abode, where they rested awhile from their fears and fatigue, partaking of some refreshment, which he had previously stored there for himself. Early in the morning he returned home to avoid suspicion; but failed not, in the course of the day, to repair again to the place which held all that was dear to him: he brought her mats to lie on, the finest *gnatoo* for a change of dress, the best of food for her support, sandal-wood oil, cocoa-nuts, and every thing he could think of to render her life as comfortable as possible. He gave her as much of his company as prudence would allow, and at the most appropriate times, lest the prying eye of curiosity should find out his retreat.

But, though happy in each other's affections, during their sojourn in this secluded cave, the length of time he found it necessary to be absent from his bride, to prevent suspicion and detection, was a great source of discomfort; and he longed for an opportunity to arrive, when he might without hazard acknowledge her as his chosen wife, and restore her to liberty and security. At length he proposed to his vassals an emigration to the Feejee Islands, and requested them to accompany him. They complied, but asked him respectfully, if he would not take a Tonga wife with him. He laughingly replied, no; but that he might possibly find one by the way. Having put to sea, he steered by the cliffs of Hoonga, the isle of the

grotto; and suddenly bidding his crew wait while
he fetched his wife, dived, to their astonishment,
beneath the wave. They waited awhile in the
greatest suspense and wonder; and at length, when
they had despaired of seeing him more, how was
their astonishment increased to see him suddenly
appear, accompanied by a lovely female! Soon,
however, they recognized her features as those of
one whom they had believed to have been slain,
in the general massacre of her family; but having
been briefly informed by the chief of the events
that had transpired, they joyfully congratulated him
on his happiness. At length they arrived safely
at Feejee, where they resided under the protection
of a chief two years; when, hearing of the death
of the tyrant from whose persecutions they had fled,
the young chief returned with his wife to their
native island, and lived long in peace and happiness.

The only point of difficulty in this pleasing story
is the time which the young bride is said to have
spent in the cavern; viz., two or three months; as
it is not easy to understand how the air could have
remained so long fit for the support of life, if unrenewed by communication with the atmosphere.
However, it is quite probable, that there might
have been clefts in the ceiling, which might admit
air without admitting light; although Mr. Mariner
could discover none, even by swimming up each
of the chambers with the torch in his hand. He,
however, bears testimony, expressly, to the purity
of the air during his visit to the retreat, so that
we will not reject the narrative on that account.

The islands of the third class differ greatly in
appearance and structure from those of either of
the preceding. Abundant traces of their volcanic
origin show that they have been elevated from the
bed of the Ocean by the resistless energy of fire,
which has given a bold and irregular form to their
rocky mountains that greatly increases the romantic

VOLCANIC ISLANDS.

beauty of their scenery. Every visitor to the South
Seas has spoken in eulogy of these lovely islands.
The highly-wrought descriptions given in Cook's
voyages are declared by recent writers to be no
whit beyond the reality. Instead of the long, low
coral island, with its grove of cocoa-nut trees almost

springing from the water's edge, these islands rise
up from the sea in tall cliffs, or gentle slopes, while
the towering mountains of the interior, wooded to
their summits, pierce the clouds. "The mountains
frequently diverge in short ranges from the interior
towards the shore, though some rise like pyramids
with pointed summits, and others present a conical
or sugar-loaf form, while the outline of several
is regular, and almost circular." In some places
the mountain ranges terminate in abrupt precipices
frowning over the Pacific, that frets and foams below; in others, there is a broad belt of level land,
of the most fertile character, and rich in the various productions of a tropical region. To these are
now added charms of another character. When
visited by Cook, there was the loveliness and magnificence of Nature, but that was all; man was evil;
plunged in the grossest idolatry, cruelty, and licen-
tiousness, he strangely contrasted with the scenes
around him: but, now that the glad tidings of salvation through the Lord Jesus Christ have been,
by the grace of God, made known to them, how
incomparably is the scene enhanced! The wretched
hut is exchanged for the neat and picturesque cottage; cultivated fields and pleasant gardens chequer
the mountain sides; the sound of the axe and hammer has replaced the savage war-cry, and the peaceful people flock to the worship of the true God,
instead of a licentious dance before a hideous idol.
O, how far does the moral beauty of such a change
as this exceed the beauty of mere natural scenery,
though it be lovely as is that of Tahiti! Captain

Gambier has thus described his emotions on visiting these scenes:—"After passing the reef of coral which forms the harbour, astonishment and delight kept us silent for some moments, and were succeeded by a burst of unqualified approbation at the scene before us. We were in an excellent harbour, upon whose shores industry and comfort were plainly perceptible; for in every direction, white cottages, precisely English, were seen peeping from amongst the rich foliage which everywhere clothes the lowland in these islands. Upon various little elevations beyond these, were others, which gave extent and animation to the whole. The point on the left, in going in,* is low, and covered with wood, with several cottages along the shore. On the right, the high land of the interior slopes down with gentle, gradual descent, and terminates in an elevated point, which juts out into the harbour, forming two little bays. The principal and largest is to the left, viewing them from seaward; in this, and extending up the valley, the village is situated. The other, which is small, has only a few houses; but so quiet, so retired, that it seems the abode of peace and perfect content. Industry flourishes here. The chiefs take a pride in building their own houses, which are now all after the European manner; and think meanly of themselves, if they do not excel the lower classes in the arts necessary for their construction. Their wives, also, surpass their inferiors in making cloth. The queen

---

* The captain is speaking of the harbour of Fa-re, in the island of Huaheine.

and her daughter-in-law, dressed in the English fashion, received us in their neat little cottage.

"The sound of industry was music to my ears. Hammers, saws, and adzes, were heard in every direction. Houses in frame met the eye in all parts, in different stages of forwardness. Many boats, after our manner, were building, and lime burning for cement and whitewashing.

"I walked out to the point forming the division between the two bays. When I had reached it, I sat down to enjoy the sensations created by the lovely scene before me. I cannot describe it; but it possessed charms independent of the beautiful scenery and rich vegetation. The blessings of Christianity were diffused among the fine people who inhabited it; a taste for industrious employment had taken deep root; a praiseworthy emulation to excel in the arts which contribute to their welfare and comfort had seized upon all, and in consequence civilization was advancing with rapid strides."

The volcanic islands, like the first-described class, are protected from the fury of the tempestuous Ocean by the natural rampart of a coral reef. The reef is often a mile and a half, or two miles from the beach, though sometimes it approaches so close as to be connected with it, interrupting in that part the continuity of the lagoon. The usual width of the coral rock is from five to twenty or thirty yards; yet over this the waves usually break, and when rolling in upon an unbroken line of reef, perhaps two miles in length, the spectacle is one of surpassing grandeur and beauty. The

BOLABOLA.

island of Bolabola, however, is surrounded by a ring
of land almost unbroken, on which are growing
groves of cocoa-nuts; the reef being wholly elevated
above the sea.

The openings in the reefs in the larger islands
are almost invariably placed opposite the mouth
of a river. One can readily understand, that a
current of fresh water would be detrimental to the
health of a polype formed for living in the sea,
and therefore the openings here might have been
expected. But this effect is increased by the sedi-

ment deposited, as has already been observed in speaking of the coral islands. The little green wooded islets, which serve as gateways here, as in the former case, are susceptible of ready explanation. Where a river empties itself, a great quantity of vegetable matter, rubbish, and earth, is perpetually carried down, and this would naturally be deposited at the shallows on either side, where the stream met the boiling waves of the Ocean. The heap would very soon be raised, by accumulations, above the surface of the tide, decomposition would take place, seeds washed down would spring up, and, under a tropical climate, the young soil would speedily be clothed with trees and shrubs. In the small isles where there is no efflux of fresh water, the process would be more protracted, but not essentially different: the current driven in through the aperture would bring sea-weeds, and the floating matters washed off the land, and when the soil was once raised above the surface, though composed of but sand and pulverized coral, the cocoa-nut would grow and thrive. It is remarkable to see this graceful palm rising from the very sea-sand, where its roots are daily wet with salt-water, yet towering to the height of seventy feet, throwing out its elegant plumose fronds, and producing its clusters of flowers and fruit, as luxuriantly as if it were growing in the rich alluvial valleys of the interior. These little fairy islets, so useful as well as ornamental, give a very peculiar character to the prospects from the land. "Detached from the large islands, and viewed in connection with the Ocean

rolling through the channel, on the one side, or the
foaming billows dashing, and roaring, and breaking
over the reef on the other, they appear like emerald
gems of the Ocean, contrasting their solitude and
verdant beauty with the agitated element sporting in
grandeur around."

Upon the mind of a European, the sailing in a
small vessel through one of these sheltered lagoons
has a most novel and interesting effect. The shore,
on the one hand, presenting its shifting aspects
of beauty, as the boat skims past, the convol-
vulus and other brilliant creeping plants entwined
about the dark rocks, or trailing in unrestrained
wildness over the sands; the solemn groves, now
revealing their sombre and shady retreats, now projecting their massy foliage in full sun-light; the
valuable bread-fruit (*Artocarpus*), the light and
elegant aito (*Casuarina*), the magnificent tamanu
(*Callophyllum*), with its glossy evergreen leaves, the
hutu (*Barringtonia*) of giant height, adorned with
large flowers of white and pink, are relieved by the
coral-tree (*Erythrina*), with its light-green waving
leaves and bunches of scarlet blossoms, and the
hoary foliage of the candle-nut (*Aleurites*). The
cocoa-nut, always beautiful, whether growing alone
or in groves, but particularly pleasing when seen
planted around a neat white-washed cottage, in
company with the broad-leaved plantain or banana;
the light tree-ferns displaying their elegant tracery
against the sky, the native chestnut (*Tuscarpus*),
rearing its stately head above its fellows, and marking the position of a running stream;—these and

many other trees of beauty and usefulness strike the eye of a stranger. Seaward, there is the long line of the reef; a low but impregnable barrier, with the surging wave foaming over it; and, beyond, the boundless Pacific, unbroken by any object, save the white-sailed canoe in the distance, scarcely distinguishable from the crest of a wave, but perhaps freighted with the humble native missionary, bearing to some neighbouring island that gospel of Christ which he has found to be "the power of God unto [his] salvation." Beneath and around is the placid and lake-like lagoon, the progress of the boat alone dimpling its smooth face. So transparent is the water, that the varied bottom is distinctly visible many fathoms down, showing the growth of living coral branching in fantastic imitation of the shrubs and trees on the shore, and representing to the charmed imagination an extensive submarine shrubbery of many hues. Even the irregular movements of the spined urchins (*Echini*) are clearly seen as they crawl upon the sands, and the multitudes of playful little rock-fishes (*Labri*), of every rich and glowing tint, gliding with easy and graceful motion among the branches, rivet the spectator's attention.

Mr. Ellis thus describes his feeling in a similar situation, walking on the lonely sea-beach by moonlight: "The evening was fair, the moon shone brightly, and her mild beams, silvering the foliage of the shrubs that grew near the shore, and playing on the rippled and undulating wave of the Ocean, added a charm to the singularity of the prospect, and enlivened the loneliness of our situation. The

scene was unusually impressive. On one side, the
mountains of the interior, having their outline edged,
as it were, with silver from the rays of the moon,
rose in lofty magnificence, while the indistinct form,
rich and diversified verdure, of the shrubs and trees,
increased the effect of the scene. On the other
hand was the illimitable sea, rolling in solemn majesty its swelling waves over the rocks which defended the spot on which we stood. The most profound silence prevailed, and we might have fancied
that we were the only beings in existence; for no
sound was heard, excepting the gentle rustling of
the leaves of the cocoa-nut tree, as the light breeze
from the mountain swept through 'them; or the
hollow roar of the surf, and the rolling of the
foaming wave, as it broke over the distant reef,
and the splashing of the paddle of our canoe, as
it approached the shore. It was impossible, at
such a season, to behold this scene, exhibiting impressively the grandeur of creation and the insig-
nificance of man, without experiencing emotions of
adoring wonder and elevated devotion, and exclaim-
ing with the Psalmist, ' When I consider thy heavens, the work of thy fingers, the moon and the
stars which thou hast ordained; what is man, that
thou art mindful of him, or the son of man, that
thou visitest him ?' "*

The same pleasing writer has given us a vivid picture of the emotions awakened by passing a night
upon the open sea in a small boat. He was proceeding from the island of Eimeo to Huaheine:

* Polynesian Researches, 2nd ed. vol. ii. p. 245.

"Nothing can exceed the solemn stillness of a night at sea within the tropics, when the wind is light, and the water comparatively smooth. Few periods and situations amid the diversified circumstances of human life, are equally adapted to excite contemplation, or to impart more elevated conceptions of the Divine Being, and more just impressions of the insignificancy and dependence of man. In order to avoid the vertical rays of a tropical sun, and the painful effects of the reflection from the water, many of my voyages among the Georgian and Society Islands have been made during the night. At these periods I have often been involuntarily brought under the influence of a train of thought and feeling peculiar to the season and the situation, but never more powerfully so than on the present occasion.

"The night was moonless, but not dark. The stars increased in number and variety as the evening advanced, until the whole firmament was overspread with luminaries of every magnitude and brilliancy. The agitation of the sea had subsided, and the waters around us appeared to unite with the indistinct, though visible, horizon. In the heaven and the ocean, all powers of vision were lost; while the brilliant lights in the one being reflected from the surface of the other, gave a correspondence to the appearance of both, and almost forced the illusion on the mind, that our little bark was suspended in the centre of two united hemispheres.

"The perfect quietude that surrounded us was equally impressive. No objects were visible but the lamps of heaven and the luminous appearances of the deep. The silence was only broken by the murmurs of the breeze passing through our matting sails, or the dashing of the spray from the bows of our boat, excepting at times, when we heard, or fancied we heard, the blowing of a shoal of porpoises, or the more alarming sounds of a spouting whale.

"At a season such as this, when I have reflected on our actual situation, so far removed, in the event of any casualty, from human observation and assistance, and preserved from certain death only by a few feet of thin board, which my own unskilful hands had nailed together, a sense of the wakeful care of the Almighty has alone afforded composure.

"The contemplation of the heavenly bodies, although they exhibit the wisdom and majesty of God, who 'bringeth out their host by number, and calleth them all by names, by the greatness of His might,' impressed at the same time the conviction that I was far from home, and those scenes which in memory were associated with a starlight evening in the land I had left. Many of the stars which I had beheld in England were visible here: the constellations of the zodiac, the splendours of Orion, and the mild twinkling of the Pleiades, were seen; but the northern pole-star, the steady beacon of juvenile astronomical observation, the Great Bear, and much that was peculiar to a northern sky, were wanting. The effect of mental associations, connected with the appearance of the heavens, is singular and impressive. During a voyage which I subsequently made to the Sandwich Islands, many a pleasant hour was spent in watching the rising of those luminaries of heaven, which we had been accustomed to behold in our native land, but which for many years had been invisible. When the polar star rose above the horizon, and Ursa Major, with other familiar constellations, appeared, we hailed them as long-absent friends; and could not but feel that we were nearer England than when we left Tahiti, simply from beholding the stars that had enlivened our evening excursions at home."*

A stranger is forcibly struck with the remarkable fearlessness which the natives of these islands have of the sea. They appear almost as amphibious as seals, sporting about in the deep sea for many hours, sometimes for nearly a whole day together. No sooner does a ship approach a large island, than the inhabitants swim off to welcome her; and long before she begins to take in sail, she is surrounded by human beings of both sexes, apparently as much at home in the Ocean as the fishes themselves. The children are taken to the water when but a day or two old, and many are able to swim as soon as they are able to walk. In coasting along the shore, it is a rare thing to pass a group of cottages, at any hour of the day, without seeing one or more bands of children joyously playing in the sea. They have several distinct games which are played in the water, and which are followed with exceeding avidity, not only by children, but by the adult population. One of these is the fastening of a long board or pole on

* Poly. Res. iii. 164.

a sort of stage, where the rocks are abrupt, in such
a manner that it shall project far over the water:
then they chase one another along the board, each
in turn leaping from the end into the sea. They are
also fond of diving from the yard-arms or bowsprit
of a ship. But the most favourite pastime of all, and
one in which all classes and ages, and both sexes,
engage with peculiar delight, is swimming in the
surf. Mr. Ellis has seen some of the highest chiefs,
between fifty and sixty years of age, large and cor-
pulent men, engage in this game with as much
interest as children. A board about six feet long
and a foot wide, slightly thinner at the edges than
at the middle, is prepared for this amusement,
stained and polished, and preserved with great care
by being constantly oiled, and hung up in their dwell-
ings. With this in his hand, which he calls the
wave-sliding board, each native repairs to the reef,
particularly when the sea is running high, and the
surf is dashing in with more than ordinary violence,
as on such occasions the pleasure is the greater.
They choose a place where the rocks are twenty or
thirty feet under water, and shelve for a quarter of
a mile or more out to sea. The waves break at this
distance, and the whole space between it and the
shore is one mass of boiling foam. Each person
now swims, pushing his board before him, out to
sea, diving under the waves as they curl and break,
until he is arrived outside the rocks. He now
lays himself flat on his breast along his board,
and waits the approach of a huge billow; when
it comes, he adroitly balances himself on its sum-

mit, and paddling with his hands, is borne on the crest of the advancing wave, amidst the foam and spray, till within a yard or two of the shore or rocks. Then, when a stranger expects to see him the next moment dashed to death, he slides off his board, and catching it by the middle, dives seaward under the wave, and comes up behind, laughing and whooping, again to swim out as before. The utmost skill is required, in coming in, to keep the position on the top of the wave; for, if the board get too forward, the swimmer will be overturned and thrown upon the beach; and, if it fall behind, he will be buried beneath the succeeding wave; yet some of the natives are so expert as to sit, and even to stand upright upon their board, while it is thus riding in the foam.

Their sport is, however, not unfrequently disturbed by the appearance of a shark. This terrific animal is particularly abundant among the South Sea Islands, and remarkably bold and ferocious. The cry of "A Shark!" among the surf swimmers will instantly set them in the utmost terror, and generally they fly with precipitation to the shore; though sometimes they unite and endeavour to frighten him away with their shouting and splashing. Often, however, the animal is too determined lightly to give up his prey, as was the case in the following instance recorded by Mr. Richards of the Sandwich Islands:—

"At nine o'clock in the morning of June 14th, 1826, while sitting at my writing-desk, I heard a simultaneous scream from multitudes of people, 'Pau i ka mano!' (Destroyed by the shark!) The

beach was instantly lined by hundreds of persons,
and a few of the most resolute threw a large canoe
into the water, and, alike regardless of the Shark

WHITE SHARK.

THE ATTITUDE OF THE FISH IN TAKING ITS PREY.

and the high rolling surf, sprang to the relief of
their companion. It was too late; the Shark had
already seized his prey. The affecting sight was
only a few yards from my door, and while I stood
watching, a large wave almost filled the canoe, and
at the same instant a part of the mangled body was
seen at the bow of the canoe, and the Shark swimming towards it at her stern. When the swell had

rolled by, the water was too shallow for the Shark to swim. The remains, therefore, were taken into the canoe, and brought ashore. The water was so much stained by the blood, that we discovered a red tinge in all the foaming billows, as they approached the beach.

"The unhappy sufferer was an active lad about fourteen years old, who left my door only about half an hour previous to the fatal accident. I saw his mother, in the extremity of her anguish, plunge into the water, and swim towards the bloody spot, entirely forgetful of the power of her former god."*

"A number of people, perhaps a hundred, were at this time playing in the surf, which was higher than usual. Those who were nearest to the victim, heard him shriek, perceived him to strike with his right hand, and at the same instant saw a Shark seize his arm. Then followed the cry which I heard, which echoed from one end of Latraina to the other. All who were playing in the water made the utmost speed to the shore, and those who were standing on the beach saw the surf-board of the unhappy sufferer floating on the water, without any one to guide it. When the canoe reached the spot, they saw nothing but the blood with which the water was stained for a considerable distance, and by which they traced the remains whither they had been carried by the Shark or driven by the swell. The body was cut in two by the Shark, just

* The Shark was formerly worshipped in the Sandwich Islands.

above the hips; and the lower part, together with the right arm, was gone."*

A dreadful instance of the voracity of these formidable animals occurred a. few years ago among the Society Islands. Upwards of thirty natives were passing from one island to another, in a large double canoe, which consists of two canoes fastened together, side by side, by strong horizontal beams, lashed to the gunwales by cordage. Being overtaken by a storm, the canoes were torn apart, and were incapable, singly, of floating upright. In vain the crew attempted to balance them—they were every moment overturned. Their only resource was to form a hasty raft of such loose boards and spars as were in the craft, on which they hoped to drift ashore. But it happened, from the small size of their raft, and their aggregated weight, that they were so deep in the water, that the waves washed above their knees. Tossed about thus, they soon became exhausted with hunger and fatigue; when the Sharks began to collect around them, and soon had the boldness to seize one and another from the raft, who, being destitute of any weapon of defence, became an easy prey. The number and audacity of these monsters every moment increased, and the forlorn wretches were one by one torn off, until, but two or three remaining, the raft at length, lightened of its load, rose to the surface, and placed the survivors beyond the reach of their terrible assailants. The tide at length bore them to one of the islands, a melancholy remnant, to tell the sad fate of their companions.

* American Missionary Herald.

With such simple vessels as were used by these people, it is surprising that such accidents did not more frequently occur. When we consider that, before their intercourse with Europeans, they pos- sessed no metal tools, that their work was performed wholly by the eye, without line, rule, or square, and that the seams were closed merely by, as it were, *tying* the planks to each other with cinet, it does seem surprising that their canoes could even live in a sea. Yet they were strong and secure, and many of them remarkably dry and comfortable, leaking very little, for they were accustomed to insert between the seams the cocoa-nut husk, which always swells when wetted; and the expansion of this substance closed the crevices neatly. Their craft, though varying much in size and minor points, according to the purposes for which they were intended, were built nearly on the same model; the stem and stern generally being curved upwards, so as to project out of water. As they were much higher than wide, they needed some contrivance to obtain uprightness; and this they secured, either by lashing two together by cross-beams, making the double canoe just now alluded to, or by means of an *outrigger*, which is a stout plank or spar, parallel to the side of the canoe, and fixed at some distance from the larboard side, by two horizontal poles, which connect it with the vessel. The out- rigger floats on the water, and while it remains fast, there is no possibility of capsizing. They were furnished with masts, sails made of the leaves of the *pandanus*, woven into a sort of matting, and rigging made of cocoa-nut fibre, which makes good rope.

The mode in which these scattered isles were peopled is a subject of interesting discussion, as the physical character of the inhabitants, their language, and many peculiarities in their customs, seem to indicate their Asiatic origin; while, on the other hand, it was deemed highly improbable that the progress should have been made in a direction opposed to that of the trade-wind, and in such feeble craft as they possessed. But the trade-wind is occasionally exchanged for violent and continued gales in other directions; and instances have come to our knowledge, in which voyages of several hundred miles have been performed by native canoes, directly to windward. Thus, Captain Beechy found at Byam Martin Island a native of Tahiti, named Tuwarri, who, with a few companions, had sailed from Chain Island on a voyage to Tahiti; but after being out some time, he was met by a violent storm, which drove him far out of his course and knowledge. At length, after very severe privations and sufferings, he arrived at Byam Martin, four hundred and twenty miles distant in a windward direction from the point of embarkation.* Such involuntary emigrations as this, when we consider how intimately the various groups are connected with each other, and with the Indian Archipelago, seem sufficient to warrant the conclusion, that the tide of population has flowed in a direction from west to east.

* Voyage to the Pacific, &c.

In the transparent waters of the lagoons and sheltered bays, fishes of great variety and beauty are seen; and as many of them are of large size, and of exquisite flavour, the obtaining of them forms no small part of the occupation of the Polynesians. Some of their modes of fishing are highly curious and ingenious. One, which is very successful, reminds us of a wire mouse-trap. A circular space in the lagoon, of about three or four yards in diameter, is enclosed by building up a wall from the bottom to the surface, in a part where it is not very deep. In one part of the top an opening is left a foot or two wide, and five or six inches deep. From each side of this aperture another stone wall, likewise reaching to the surface, is built to the length of fifty or a hundred yards in a diverging direction, so as to include a large space of water, which is open at one end, but, becoming narrower and narrower, leads into the circular pen. Fishes are usually found in these traps every morning, which are either taken out with a hand-net, or allowed to remain till wanted, as in a preserve.

Many fishes, which have the habit of springing out of water when alarmed, are taken by means of rafts. These are from fifteen to twenty feet long, and six or eight feet wide, built of light wood, such as the native *hibiscus*. Along one side a fence or screen is raised to the height of four or five feet, by fixing a row of upright stakes in the raft, to which slender poles are attached horizontally, one above another. A large party of men proceed with twenty or thirty of these rafts to a shallow part of the lagoon, and then arrange themselves in a large circle, enclosing a large space of water. They then gradually narrow the circle by approaching each other, keeping the fenced edge of the raft on the outside. At this juncture a few persons go into the circle with a canoe, and beat the surface of the water violently with long white sticks, making as much commotion as possible. The fish, alarmed, dart away towards the rafts, and leaping out of water, endeavour to clear them; but, striking against the perpendicular fence, they fall on the raft, and are gathered into baskets, or into canoes prepared on the outside of the circle.

From the seeds of some of the native plants, a liquor is prepared, which has the property of intoxicating fishes, and rendering them insensible. The mixture is frequently poured into the water in narrow places near the shore, or upon the reef; soon after which the fish come out of their retreats, and float in considerable numbers on the surface as if dead, when they are caught without resistance.

Sometimes the long leaves of the cocoa-nut are tied up in bunches, and affixed along a line, which being carried out and dropped into the water, the two ends are towed in two canoes towards the shore. This rude apology for a net, drives many fishes into the shallows, whence they are taken out with hand-nets, or speared. Nets, however, made on the same principle as our own, are manufactured by them, and are exceedingly well made. They are of various kinds: a casting-net is used with much dexterity, being thrown from the hand over a shoal of small fishes, as the fisherman walks along the shore. Salmon-nets are made forty fathoms long, and are very effective; stones tied in bags of matting being used instead of leads, and floats of light wood for corks.

Fishing with the barbed spear is a favourite amusement in these islands. Before the introduction of iron, the implement was made of hard wood; ten or twelve pointed pieces being fastened to the end of a pole eight feet long; but now iron heads are usually employed, barbed on one side. With these spears the natives proceed to the reef, and wade into the sea as high as their waists, their feet being defended from the sharp points of the coral and the spines of the sea-urchins by sandals made of tough bark, twisted into cords. Stationing themselves near an opening in the rocks, they watch the motions of the fishes, as they shoot to and fro, and dart the spear, sometimes with one hand, but more commonly with both, frequently striking their prey with great dexterity.

The fishermen often pursue their avocation by night; sometimes in the dark, sometimes by moonlight, but more usually by torchlight. Their torches are either large bunches of dried reeds firmly tied together, or else are made of the candle-nut (*Aleurites triloba*), which the natives use to light their houses. These nuts are heart-shaped, about as large as a walnut, and enclosed in a very hard shell. After being slightly baked in an oven, the shell is removed, a hole bored through the kernel, and a rush passed through the hole, when they are hung up in strings for use. Torches are made by enclosing four or five strings of the nuts in the leaves of the screw-pine (*Pandanus*), which not only keep them together, but increase the brilliancy of the light.

These nocturnal fishing expeditions are described as producing a most picturesque effect. Large parties of men proceed to the reef, when the sea is comparatively smooth, and hunt the totara, or hedge-hog-fish, probably a species of *Diodon:* and it is a beautiful and interesting spectacle, to behold a long line of reef illuminated by the flaming torches, the light from which glares redly upon the foaming surf without, and the calm lagoon within. Each fisherman holds his torch in his left hand high above his head, while he poises his spear in his right, and stands with statue-like stillness, watching the approach of the fish.

A similiar mode of fishing is practised in the rivers, and though the circumstances are different, the effect is not inferior. "Few scenes," says Mr. Ellis, "present a more striking and singular effect, than a band of natives walking along the shallow parts of the rocky sides of a river, elevating a torch with one hand, and a spear in the other; while the glare of their torches is thrown upon the overhanging boughs, and reflected from the agitated surface of the stream; their own bronze-coloured and lightly-clothed forms, partially illuminated, standing like figures in relief; while the whole scene appears in bright contrast with the dark and almost midnight gloom that envelopes every other object."*

Another mode of fishing by torchlight is described by the late Mr. Williams, who accompanied some natives of Atiu on an excursion. The object of the pursuit was the Flying-fish, which is only taken by night. Double canoes were used, which, having been dragged from the rocks, thirty feet above the level of the water, down a broad sloping ladder, were launched over the surf. A torch was lighted, and the principal fisherman took his station on the fore

FISHING BY TORCHLIGHT.

part of the canoe, bearing a ring-net attached to a light pole twelve or fifteen feet long. The rowers

* Poly. Res. i. 150.

now commenced paddling with all their might, while
the headsman produced a great noise by stamping on
the hollow box of the canoe. The Flying-fish, which
were securely feeding at the outer edge of the reef,
terrified by the noise and splashing of the oars,
darted out to sea. The torch answered a double purpose; enabling the headsman to discern his prey, and
dazzle the eyes of the fishes; and as they dashed past
the canoe, on the surface of the water, he thrust
forward his net, and turned it over upon them.
Many of the natives have acquired great skill in
this exercise, and the quickness of their sight, and
the celerity of their movements are astonishing; so
that sometimes vast quantities of fish are taken in
this manner.*

A large number of fishes are taken with the hook,
as by more cultivated nations; and with all the
superiority in art, and all the advantage of metals
possessed by Europeans, the native-made hooks are
preferred, as far more effective than ours. Many
of them are really beautiful productions, and, when
we consider their total want of metallic tools, excite our astonishment at the skill and ingenuity of
the manufacturers. Our hooks are all made on one
pattern, however varying in size; but the forms
of theirs are exceedingly various, and made of different substances, viz., wood, shell, and bone. "The
hooks made with wood are curious; some are exceedingly small, not more than two or three inches
in length, but remarkably strong; others are large.
The wooden hooks are never barbed, but simply

* Missionary Enterprises, p. 270.

pointed, usually curved inwards at the point, but sometimes standing out very wide, occasionally armed at the point with a piece of bone. The best are hooks ingeniously made with the small roots of the aito-tree, or iron-wood (*Casuarina*). In selecting a root for this purpose, they choose one partially exposed, and growing by the side of a bank, preferring such as are free from knots and other excrescences. The root is twisted into the shape they wish the future hook to assume, and

POLYNESIAN FISHING-TACKLE.

allowed to grow till it has reached a size large enough to allow of the outside or soft parts being removed, and a sufficiency remaining to form the hook. Some hooks thus prepared are not much thicker than a quill, and perhaps three or four inches in length. Those used in taking sharks are formidable-looking weapons; some are a foot or fifteen inches long, exclusive of the curvatures, and not less than an inch in diameter. They are such frightful things, that no fish, less voracious than a shark, would approach them. In some the marks of the sharks' teeth are numerous and deep, and indicate the effect with which they have been used."*

The most curious, as well as most serviceable hooks, are made of the inner part of the shell of the pearl-oyster, or other large bivalves, the interior of which is pearly, called mother-of-pearl. These have great care and pains bestowed upon them: the smaller ones are cut almost circular, and made to resemble a worm, thus answering the purpose of bait as well as hook. A much larger kind is that used for the capture of the albacore, bonito, and coryphene. The shank is about six inches in length, and nearly an inch in width, cut out of pearl-shell, in the shape of a small fish, and finely polished. The barb is formed separately; it is an inch and a half in length, and is firmly bound in its place by a bandage of fine flax. The line is fastened to this, and braided all along the curve of the hook, and again fastened at the head. Sometimes a number of long bristles are attached to the shell to mimic the appearance of the Flying-fish.

The line is affixed to the end of a long bamboo rod; and the anglers, sitting in the stern of a light

* Ellis.

single canoe, are rowed briskly over the waves. The
rod is held so that the hook shall just skim the
tops of the billows; the albacore or bonito, deceived
by the resemblance, leaps after the fancied Flying-
fish, and finds itself a prey. Twenty or thirty large
fishes are occasionally taken by two men in this
manner, in the course of a morning.

A still more ingenious mode of deception is practised upon these large fishes, by employing a swift
double canoe, from the bows of which projects into
the air a long curved pole resembling a crane. At
some distance from the end this divides into two

ANGLING IN A DOUBLE CANOE.

branches, which diverge from each other. The foot
is secured in a sort of socket between the two canoes,
and is so managed that the ends of the pole are

2 D

capable of being lowered or elevated by a rope which proceeds from the fork. A man sits in the high stern, holding this rope in his hand, and watching the capture of the fishes. From the end of the projecting arms depends the line, with the pearl-hook fashioned to resemble the Flying-fish. To increase the deception, bunches of feathers are fastened to the tips of the arms, to represent those aquatic birds which habitually follow the Flying-fish in its course, to seize it in the air. The presence of these birds is so sure an indication of the position of the fish, that the fishermen hasten to the spot where they are seen hovering in the air. The canoe skims rapidly along, rising and falling on the waves, by which a similar motion is communicated to the hook, which skips along, sometimes out and sometimes in the water, while the plumes of feathers flutter immediately above. The artifice rarely fails to succeed; if the bonito perceives the hook, he instantly engages in pursuit, and if he misses his grasp, perseveres until he has seized it. The moment the man in the stern perceives the capture, he hoists the crane, and the fish is dragged in, and thrown into a sort of long basket, suspended between the two canoes. The crane is then lowered again, and all is ready for another candidate.

Yet another mode of fishing, not wanting in ingenuity, is adopted by the inhabitants of the Samoa group. A number of hollow floats, about eight inches in height, and the same in diameter, are attached to a stout cord, a short distance apart. To each of them a line is attached, about a foot in length, to the end of which a piece of fish-bone is suspended by the middle. This bone is ground exceedingly sharp at each end, so that when it is seized by the fish, the points enter the mouth in contrary directions, and secure it. The floats answer other purposes besides the obvious one of regulating the depth of the snare, attracting the fish by the whiteness of their surface, and showing by their motion when the prey was taken.

Not only in the smooth waters of the lagoon channels is the hook and line used, but in the open Ocean; as, notwithstanding the frail character of their vessels, the barbarous natives of these oceanic isles are skilful and fearless in navigation. Even the terrific shark is attacked in his own element; sometimes involved in a net, when frequently he makes havoc among the fishermen before he can be transfixed by their spears; and sometimes caught, as intimated above, with the insidious hook. The most daring young men, usually the chiefs, are the first to assault the monster; while the elders watch the proceedings in their canoes from a distance, partakers of the excitement, though no longer sharers of the heroism. The eagerness with which these expeditions are set on foot, and the ardour with which they are prosecuted, are only equalled by the excited feelings of those who, in other countries, pursue the more noble objects of the chase.

The fishes of these seas are, many of them, interesting; some of them have been already named. The Albacore and the Bonito are common in the tropical parts of the Pacific, and are both members of the Mackerel family. They are of considerable size, but the Albacore (*Scober Germo*) is the larger, sometimes being found six feet in length. Like its relative, our own Mackerel, it is a fish of much elegance, and its colours are beautiful. The back is bright azure, with a golden tint; the belly and sides silvery, with rainbow reflections, like mother-of-pearl, and the same notched fins near the tail are bright yellow. In slight winds, when the motion of a ship is slow, these fishes are usually to be seen around her; if she be becalmed, and consequently motionless, they remain at some little distance, when the most tempting bait is ineffectual; but if she be sailing rapidly before a brisk breeze, they pertinaciously keep her company, keeping close alongside, and seizing the hook with avidity. The Albacore, as already hinted, is one of the hunters of the little Flying-fish. It is said to be highly interesting to watch one of these fishes keenly engaged in pursuit of its volatile prey: to mark the precision with which it keeps exactly beneath during the aerial leaps of the victim, keeping it steadily in sight, prepared to snap it up, on the instant of its submersion. The Flying-fish, however, by its exceeding agility, darting again into the air in a moment, sometimes contrives to escape the fearful jaws of its adversary.

The Albacore, in its turn, has occasion to exercise cunning and contrivance, to evade the attacks of a still mightier foe. Mr. F. D. Bennett mentions that, on one occasion, "The Albacore around the ship afforded us an extraordinary spectacle; they were collected close to the keel of the vessel, in one dense mass, of extraordinary depth and breadth, and swam with an appearance of trepidation and watchfulness. The cause of this unusual commotion was visible in a Sword-fish lurking astern, awaiting a favourable opportunity to rush upon his prey when they should be unconscious of danger, or away from the protection of the ship. The assembled Albacore continued, in the mean time, to pass under the keel of the vessel, from one side to the other, often turning simultaneously on their side to look for the enemy: their abdomens glittering in the sun as a wide expanse of dazzling silver. It was evident that the Sword-fish desired but a clear field for his exertions; and in the course of the day we observed him make several dashes amongst the shoal, with a velocity which produced a loud rushing sound in the sea; his body, which, when tranquil, was of a dull brown colour, assuming at these times an azure hue." *

Mr. Bennett conjectures with much probability, that it is as a protection against the attacks of the Sword-fish, that Albacore and other fishes so often attach themselves to a ship, or the body of a whale; the vicinity of so large a body being sufficient to deter the former from making his impetuous thrusts among the shoal, lest his bony weapon being driven into the solid substance by the violence of his assault, he might not be able to retract it. Instances are not rare, however, in which the Sword-fish, perhaps forgetting his usual caution, (for he is re-

* Whaling Voyage, vol. i. p. 270.

2 D 2

puted a very cautious fish,) has left his sword in
the hull of a ship. The Foxhound, a South Sea
whaler, was cruising in the Pacific in 1817, when
one day, when most of the crew were below at
dinner, a loud splashing was suddenly heard in the
sea by a New Zealander on deck, who, on looking
over the side, saw a large dark body sinking, and
immediately gave the alarm of a man overboard.
The crew, however, were found to be complete,
and the occurrence passed over. Soon after, one
of the men observed a rugged object projecting
from the vessel's side, which, on examination, proved
to be the snout of a Sword-fish, with part of the
head attached, broken off by the fracture of the
skull. On the vessel's arriving at Sydney, the projecting part was sawn off, after vain endeavours to
extract the weapon; and at the conclusion of the
voyage, the pierced wood was taken out and placed
in the British Museum.

It is worthy of observation that, with very few
exceptions, the immense population of the Ocean
is carnivorous. The principal circumstance that
regulates the choice of diet among fishes seems to
be the power of mastery. Of terrestrial creatures,
a very large number are peaceful, never, under
ordinary circumstances, willingly taking the life of
even the most helpless around them; but the sea
is a vast slaughter-house, where nearly every inhabitant dies a violent death, and finds a grave in the
maw of his fellow. We have just seen the Sword-
fish preying upon the Albacore, and the Albacore
upon the Flying-fish; while the Flying-fish itself,

though so general a favourite, is the greedy devourer of other fishes smaller than itself. Yet let us not arraign the providence of God, as if it were cruel and unkind: a sudden termination of existence is the most merciful mode, as far as we can conceive, by which the overflow of animal life could be checked.

"Harsh seems the ordinance, that life by life
  Should be sustain'd; and yet when all must die,
And be like water spilt upon the ground,
Which none can gather up,—the speediest fate,
Though violent and terrible, is best.
O, with what horrors would creation groan,
What agonies would ever be before us,—
Famine and pestilence, disease, despair,
Anguish and pain in every hideous shape,
Had all to wait the slow decay of Nature!
Life were a martyrdom of sympathy;
Death, lingering, raging, writhing, shrieking torture;
The grave would be abolished; this gay world
A valley of dry bones, a Golgotha,
In which the living stumbled o'er the dead,
Till they could fall no more, and blind perdition
Swept frail mortality away forever.
'Twas wisdom, mercy, goodness that ordain'd
Life in such infinite profusion,—Death
So sure, so prompt, so multiform to those
That never sinn'd, that know not guilt, that fear
No wrath to come, and have no heaven to lose."*

Before we leave these charming regions, we will for a moment notice a few other of the various tribes of living beings that make the sea their home. A curious example of instinctive stratagem occurs in a little crab (*Hyas*——?) which is common upon the shore-reefs. It is about six inches in circum-

* Pelican Island.

ference, of a dull brown hue, the body and legs entirely covered with stiff, curved bristles. It covers itself with decaying vegetable rubbish, mud, sand, &c., and thus lies in ambush for its passing prey. Thus masked, it maintains its assumed character by the most sluggish movements, as if the little heap were slightly moved by the tide; but, when taken into the hand, or otherwise alarmed, it can be sufficiently active. The spines upon its body to retain the rubbish, the short but strong claws easily concealed, the eyes placed at the end of long footstalks, curving upwards and thus raised above the mass, show beautiful adaptations of its structure to its economy.

Another crab of the reef (*Calappa tuberculata*), makes use of another artifice for concealment. It is heart-shaped, with the margin of its shell projecting broadly. When alarmed, it draws its feet under the margin, and folds them close to its side, claps its broad flat claws upon its head, and lies motionless, in which state it may be handled without manifesting any sign of life. A sailor seeing one of these little crabs on the shore, picked it up, and after admiring it awhile, put it into his pocket as a "curious stone;" he was presently astonished by the efforts of his prize to escape from durance vile.

On the barrier reefs are found elegant animal-flowers (*Diazona*), expanding their numerous tentacles of pink and white, which form a wide circular disk, at the summit of a round fleshy stem. If touched, or otherwise alarmed, they rapidly fold inwards their beautiful tentacles, and sink to the rock, contracting to a very diminutive size, so as easily to elude observation. The same reefs are enlivened also by numbers of another species of Sea-anemone (*Zoanthus*), which cover large surfaces of the rock, like beautiful carpets or mats of wide expanse. When opened beneath the water, under the beams of the sun, they display a series of squares with elevated margins, the interior being of a bright green, the exterior of a fawn colour. These, also, contract instantly on the slightest touch; and thus entire fields of them, being connected together by a common fleshy disk upon the rock, are changed in a moment, as if by magic, from brilliant green to dull brown, which again, as they recover from their alarm, is soon replaced by the verdant hue.

Numerous species of Squid and Cuttle are observed in the Pacific, several of which have the power of making long leaps out of the water, even to the same height and distance as the Flying-fish, whence these kinds are denominated by seamen, Flying Squid. One of these, which appears to have been an *Onychoteuthis*, is described by Mr. F. D. Bennett, as having fallen, in one of its leaps, upon the deck of the ship in which he was sailing. The whole class to which these animals belong is remarkable for the powerful apparatus with which the animals are endowed for seizing prey, in the numerous long and flexible arms, furnished with cup-like suckers, which forcibly adhere to any object at the will of the creature. But the genus just mentioned is favoured above its fellows; for, in addition to the usual structure, there is placed in each

21

sucker-cup of the long feet, a sharp projecting hook. On the smooth and glossy scales of fishes, lubricated with slime, it might not be always easy at once to create a vacuum; but these hooks are plunged by the action of the sucker into the flesh of the struggling victim, whereby a firm hold is obtained, and the prey is dragged to the powerful beak.

Some of these animals frequent the crevices and holes of the rocks, whence they protrude their long arms for the capture of prey. They form an acceptable article of food to the South-Sea islanders, who have exercised their ingenuity in devising a mode of entrapping them. The instrument employed for this purpose is described as a straight piece of hard wood, a foot long, round, and polished, and not half an inch in diameter. Near one end of this, a number of the most beautiful pieces of the cowry, or tiger-shell, are fastened one over another, like the scales of a fish or the plates of a piece of armour, until it is about the size of a turkey's egg, and resembles the cowry. It is suspended in a horizontal position, by a strong line, and lowered by the fisherman from a small canoe, until it nearly reaches the bottom. The fisherman then gently jerks the line, causing the shell to move as if inhabited by an animal. The Cuttle, deceived by the appearance of the supposed cowry (for no bait is used), darts out one of its arms, which it winds around the shell, adhering fast by its suckers. The fisherman continues jerking the line, and the Cuttle strengthens its hold by affixing more of its arms, until its adhesion is very strong, when, rather than quit its prey, it permits itself to be dragged from its retreat to the surface of the water, and captured.*

There are certain species of oceanic birds which it is difficult to identify with any particular region, as they are true cosmopolites. The Tropic-birds, Albatrosses, Terns, Petrels, and Boobies, are of this extended character, following and attending the voyager for many thousands of miles, and even from one ocean into another. Yet there are certain, though somewhat indefinite, limits to their range; limits governed, however, by climate, rather than by physical boundaries. Thus the Dusky Albatross (*Diomedea fuliginosa*) was observed by Captain Beechy to be numerous in the Atlantic from the Rio de la Plata to the latitude of 51° south; when it suddenly disappeared; but after rounding Cape Horn, the species again occurred at the very same latitude of 51°, and continued numerous all up the coast of Chili.

The Tropic-birds (*Phaeton*) in like manner, as their name imports, chiefly frequent the Ocean within the tropics; and according to the statements of all voyagers, are very rarely seen beyond the parallel of 35°. In a voyage to Newfoundland, however, in 1827, I frequently saw the Tropic-bird, though our latitude during the whole voyage was not so low as 40°. Elevated in the air, far above the mast-head, the long projecting tail-feathers, looking like a single slender shaft, while it turns its head to

* Ellis.

and fro, as on suspended wing it examines the vessel below, it is not liable to be confounded with
any other ocean-bird. The seamen have given it
the name of "boatswain;" perhaps on account of its
shrill whistling note, like the official call of that
authoritative personage; or, as I was told, because
it carries a *marline-spike*. This was, doubtless, *P.*
Ætherius; which has the feathers of the tail white,
but the Pacific species (*P. Phœnicurus*) is much
more handsome, the tail being scarlet. They are
thoroughly ocean-birds, rarely approaching the land
except to lay and hatch their eggs. The Red-tailed
Phaeton excavates a hollow in the sand for this
purpose, beneath the shade of bushes, where she
lays one egg: the islanders frequently take the old
birds from the nest, for the tail-feathers, which are
highly esteemed.

The Albatrosses are large birds, being but little
inferior to a swan in size. The floating carcass of a
whale affords a rich feast to many sea-birds, among
which these are pre-eminent, now swooping in the
air, now alighting on the body, now swimming and
feeding on the fragments of oily fat that escape;
now screaming harshly as they quarrel for the offal.
They are powerfully endued for flight, and make
vast excursions from land, ranging through the whole
Atlantic and Pacific Oceans.

I have already alluded to the singular manner in
which the body of a sea-bird is penetrated by air.
Mr. Bennett records a very curious circumstance
resulting from this structure, in the case of a bird
allied to the Albatross, taken in the Pacific Ocean.

It "was shot in the wing, and brought on board alive, fighting savagely with its beak and feet. With a view to preserving its plumage uninjured, I endeavoured to destroy the bird by compressing its windpipe; but found that as the breathing became laborious, a loud whistling sound was emitted from some part of the body; and upon close investigation traced it to the bone of the wing, which was fractured across, and projected through the skin, and admitted within its tube a forcible current of air, whenever the lungs made an effort at respiration: the bird was, in fact, breathing through its broken wing; and so sufficient was the supply of air the lungs received through this novel channel, that I was wearied by my attempts to suffocate my prize, and was compelled to destroy it in another manner."*

Every one who has read the romantic narratives of the old voyagers, is familiar with the name of the Booby (*Sula fusca*), so named by seamen from its apparent stupidity and familiarity, suffering itself to be knocked down with a stick or taken with the hand, when it alights, as it often does, on the spars or shrouds of a vessel. This habit seems quite unaccountable; many other birds have manifested a similar fearlessness of man when first discovered, but have soon learned the necessity of precaution: but the Booby will manifest the same unnatural tameness after being long accustomed to the cruelty of man. It does not arise from helplessness, as it is a bird of powerful wing, like its relative, the com-

mon Gannet; neither is it a sufficient explanation
to affirm, as is sometimes done, that it arises from
a peculiar difficulty in rising to flight after alighting, because it is not unfrequently caught in the air
by the hand; so incautiously does it approach man.
Notwithstanding this apparent stupidity, the Booby
is a dexterous fisher: hovering over a shoal of fishes,
he eagerly watches their motions, turning his head
from side to side in a very ludicrous manner; he
presently sees one of the unwary group approach
the surface; down he pounces like a stone, plunging
into the wave, which boils into foam with the shock.
Nor fails he to seize the scaly victim, with which
he emerges into the air, and soon it is lodged
whole in his capacious stomach. But the Frigate-
bird (*Tachypetes aquilus*) has watched the proceeding,
and instantly betakes himself to the pursuit; flight
is vain from the swiftest ranger of the Ocean, whose
extended wings measure a width of seven feet. The
Frigate-bird swooping down upon the unfortunate
Booby, compels him to disgorge the fish which he
has just swallowed, and which, long ere it can reach
the water, is seized, and again devoured by the op-
pressor.

The Frigate-bird neither swims nor dives; the
seamen fully believe that it even sleeps upon the
wing; whether this be so or not, there is good
evidence that the same individuals will remain in
the air for several successive days: they are never
known to alight on a vessel. Though the chase of
the Booby is so usual as to be considered one of
its constant means of dependence, yet it also fishes

for itself; precluded, however, from plunging into the sea; it can take only such as, like the Flying-fish, leap into another element. With such success, however, does it attack these, that it has been seen to snap up three in succession in the course of a few minutes. If, after having captured a fish, it is awkwardly placed in the beak, it hesitates not to drop it, secure of seizing it again in the descent.

To the immense congregations of aquatic birds, for the purpose of hatching and rearing their young in places congenial to their habits, allusion has already been made; and the following picture, vividly drawn by the pen of an accomplished naturalist, is probably not overcharged.

Le Vaillant, on visiting the tomb of a Danish captain at Saldanha Bay, near the Cape of Good Hope, beheld, after wading through the surf, and clambering up the rocks, such a spectacle as he supposed had never appeared to the eye of mortal. "All of a sudden, there arose from the whole surface of the island an impenetrable cloud, which formed, at the distance of forty feet above our heads, an immense canopy, or rather a sky, composed of birds of every species, and of all colours;—cormorants, sea-gulls, sea-swallows, pelicans, and, I believe, the whole winged tribe of that part of Africa, was here assembled. All their voices mingled together, and, modified according to their different kinds, formed such a horrid music, that I was every moment obliged to cover my head to give a little relief to my ears. The alarm which we spread was so much the more general among these innumerable regions of birds, as we principally disturbed the females which were then sitting. They had nests, eggs, and young to defend. They were like furious harpies let loose against us, and their cries rendered us almost deaf. They often flew so near us, that they flapped their wings in our faces, and though we fired our pieces repeatedly, we were not able to frighten them: it seemed almost impossible to disperse this cloud."

THE INDIAN OCEAN.

THE remaining great division of the waters of our globe is considerably less extensive than either of the others, but is scarcely less important, inasmuch as it is the pathway of the richest commerce of the world, the high road on which are borne the gems, and gold, and spices of the gorgeous East. It is separated from the Pacific by that grand assemblage of islands known as the Oriental Archipelago, which, for their immense size, the teeming luxuriance of their vegetation, and the valuable character of many of their productions, have no rivals. The isles of New Guinea, Borneo, and Sumatra are the largest in the world: their soil possesses a fertility that seems inexhaustible; their produce consists of the nutmeg, the clove, and other costly spices; frankincense, camphor, and other odoriferous gums; diamonds, rubies, and other precious stones; gold, silver, silks, tortoise-shell, pearls, sandal-wood, and drugs, the most valued of earthly things.

It is a singular fact, that at the very same point of time when the genius and daring of Columbus were leading Spain into the possession of a new world in the west, Portuguese enterprise was laying open the still more splendid and gorgeous regions of Asia in the east. It was in 1497 that Vasco de

2 E 2   (329)

Gama rounded the Cape of Good Hope, and penetrated to climes which had hitherto been invested
with all the romance of mystery and fable; then
commencing a commerce which has poured incalculable wealth into the lap of Europe.

This immense archipelago, which occupies a tract
of the Ocean four thousand miles in length, and
fourteen hundred in breadth, is an assemblage of
islands perfectly unique. The multitudinous islets
of the Pacific, if all united, would not together form
a third-rate island of this group. The land, though
broken with countless thousands of isles, so equally
divides the space with the sea, that one is at a loss
to say which predominates. A large majority of
the smaller isles and reefs are of the same structure as the coral atolls of Polynesia, and present
a similar character in their zoology and botany; but
the larger tracts of land, almost a continent in their
dimensions, are of the old formations. The widely-
scattered groups of small islands on the northern
boundary, indeed,—the Ladrones, the Carolines, the
Pelews, &c., we are at a loss to distinguish: they
are usually arranged in the Indian Archipelago,
while they are decidedly Polynesian in their characters.

The boats which are used by the natives of these
islands, from their very peculiar construction, as
well as from their unrivalled powers of sailing,
demand a moment's notice. Lord Anson, who first
met with them at the Ladrone Islands, and who calls
them flying proas, considers them "so singular and
extraordinary an invention, that it would do honour

to any nation, however dexterous and acute. Since, if we consider the aptitude of this proa to the navigation of these islands, which, lying all of them nearly under the same meridian, and within the limits of the trade-wind, requires the vessels made use of in passing from one to the other to be peculiarly fitted for sailing with the wind upon the beam; or, if we examine the uncommon simplicity and ingenuity of its fabric and contrivance, or the extraordinary velocity with which it moves, we shall in each of these particulars find it worthy of our admiration, and deserving a place amongst the mechanical productions of the most civilized nations, where arts and sciences have most eminently flourished."*

In direct contradiction to the practice of civilized nations, the proa is built with the two ends alike, but the two sides different. It is intended never to turn, but always to present the same side to the wind; the bow becoming the stern, and the stern the bow, at pleasure. The ends of the boat are high and project much above the water; the windward side is rounded, as in other vessels; but the lee side is flat, and almost perpendicular. As the depth greatly exceeds the breadth, it would, of course, instantly fall over on the leeward side, but for an ingenious contrivance already alluded to as used in the Polynesian islands. A light but strong frame is run out horizontally to windward, to the end of which is fastened a hollow log, fashioned into the shape of a small boat, which floats upon the

* Anson's Voyage, p. 339.

water, preventing the capsizing of the proa in that
direction; while the weight of the apparatus, called
an outrigger, prevents the same accident on the
other.   A mast rises perpendicularly from the windward edge of the proa, fastened to the heel of the
outrigger; a bamboo yard is slung near the mast-

PROAS OF THE LADRONES.

head, so that its foot shall come into the boat in
a diagonal direction near the head, there being a
socket at each end to receive the foot of the yard,
according as the proa is on either tack.   The sail
attached is made of matting, and is triangular, the
lower side being fastened to a boom running horizontally from the foot of the yard over the stern.
When it is intended to alter the course by going
upon another tack, the foot of the yard is lifted
from the one socket, carried round to leeward, and

placed in the other, while the fast sheet being let
fly, and the loose sheet hauled in, the boat is
immediately trimmed again, without loss by lee-way.
From their extraordinary power of lying near the
wind, that is, of sailing nearly towards the point
from which the wind is blowing, as well as from
their extreme narrowness cutting the water with
little resistance, these boats are the fleetest vessels
known. Anson affirms that they will run nearly
twenty miles an hour, which, though greatly short
of what the Spaniards report of them, is yet a prodigious degree of swiftness. In more modern voyages, we find the native boats called by the names
of *prows* and *prahus;* as they seem, however, to
refer to vessels of the same construction as those
described by Anson, they are probably to be considered as somewhat closer approximations to the
true pronunciation of the native name.

The navigation of these seas is rendered peculiarly unsafe by the swarms of Malay pirates by
which they are infested. Voyagers continually allude
to the alarm which every collection of native boats
inspires, as being so exceedingly swift, and the
men merciless and daring. Whole colonies of these
desperate adventurers proceed from Magindanao to
the coast of Borneo, where they seek some convenient, but retired, harbour, in which they make
their home; not living, however, upon the land,
but on board their *prahus* (or proas), which are frequently of sixty tons' burthen. During the south-
east monsoon they cruise about near the entrance
of the Straits of Malacca, ready to pounce upon

the native traders resorting to Singapore; when
about to return home, they surprise some defenceless native village, and carry off the whole of the
inhabitants to be sold into slavery. During the
absence of the pirates, their wives and children
remain in the harbour, to take charge of the booty
that may be brought in; and as these are scarcely
less warlike than the men, no other guard is necessary against the inoffensive natives of Borneo. When
the band has acquired a considerable amount of
plunder, they return to their own island, and others
supply their place. Even in the neighbourhood of
Singapore, although a British dependency, the Malay pirates absolutely swarm. The numberless little
islands in the Straits, divided by channels known
only to themselves, are like so many impregnable
fastnesses, into which they drag their unfortunate
victims, and plunder them at their leisure, defying
pursuit. The occupation has acquired all the form
and regularity of a system. A chief of some petty
Malay state, whose fortunes have been rendered
desperate by gambling, collects around him a few
adventurous and restless spirits, and sails to some
retired island. A village is formed, as a depôt for
the booty, and the armed prahus lie in wait or prowl
about. If the adventure prove successful, the chief
soon gains accessions; the village grows into a town;
and the fleet separates into squadrons, which scour
the seas of different localities. They usually sail
in company, the fleets consisting of three to twenty
prahus, each of which carries large and small guns,
and from fifteen to forty men. The captured vessels

are burnt at the depôt, and the goods put on
board prahus disguised like traders, and sold at
Singapore. The captives are sold into slavery at
Sumatra, to work on the pepper plantations of the
Malays.

Though their assaults are generally upon the
native trading-boats, yet occasionally they venture to
attack square-rigged craft.

"An English merchant, who had resided several
years in Java, embarked at Batavia on board one
of his own vessels, a large brig, taking with him
a considerable sum of money for the purchase of
the produce of the eastern districts. These facts
having reached the ears of a famous piratical chief,
he determined to waylay the vessel, and accordingly
mustering a sufficient number of prahus, cruised
about, and meeting with the brig as he had expected, commenced an attack upon her. The crew of
the latter vessel consisted of two Englishmen, the
captain and the chief officer, and about thirty Javanese seamen, who, together with the owner, defended
the vessel for some time. Towards the evening,
however, the unfortunate merchant was killed by a
spear fired from a musket, and the pirates taking
advantage of the confusion produced by this event,
immediately boarded. The two remaining Englishmen, being well aware that certain death awaited
them should they remain, threw themselves into the
sea, and succeeded in reaching a bamboo fishing-
buoy. The pirates, too busily employed in plunder-
ing their prize to think of any thing else, did not
perceive their place of refuge, and the vessels soon

drifted away out of sight. The condition of the
persons who had thus escaped had altered very little
for the better; they were immersed to the neck in
water, dreading every moment the attack of sharks:
nor had either, during the whole of the night, the
comfort of knowing that his companion was still in
existence. Soon after daylight some fishermen appeared, by whom they were perceived; but instead
of rescuing them immediately from their perilous
situation, the Javanese consulted together for a few
minutes, and then approached the sufferers, and
demanded who they were. On being told they were
Englishmen, whose vessel had been attacked and
captured by pirates, they were taken on board,
treated kindly, and conveyed to the Dutch Settlement
at Indramayo. Had they belonged to one of the
Dutch cruisers, their fate would probably have been
different; for the fishermen are on bad terms with
the officers of the government prahus, whom they
accuse of robbing them of their fish."*

The pirates who thus infest the Indian Archipelago are invariably Mahometans; none of the Pagan
natives ever being known to engage in these mur-
derous expeditions. They show no mercy: the
Europeans that fall into their hands are murdered,
and the native seamen sold into slavery.

The larger islands of the archipelago do not present a very interesting appearance from the sea.
Though clothed from the tops of the mountains
down to the very water's edge with the most luxuriant vegetation, it is too uniform to be agreeable.

The eye seeks in vain for some variation, some break' in the vast forest; all is rich massy foliage, like enormous heaps of green velvet. The solemn silence that prevails, joined with this gorgeous uniformity, creates an oppressive feeling of awe and loneliness. And when the dews of evening descend, and the gentle breeze blows off the land, it comes loaded with what have been described as spicy odours, but which are, in sober reality, but the sickly sweats produced by immense masses of vegetation in decomposition. They bear, in fact, the pestilence upon their wings.

But while this is the general character of the great islands, there are exceptions. Java, settled by the Dutch, contrasts with Sumatra and Borneo; the gloom of the forest is enlivened here and there by verdant fields and lawns, while the white villas of the Europeans chequer the hills, and give a peaceful and inviting air to the landscape. The smaller isles are said to be exquisitely lovely.

"The sea near Batavia is covered with innumerable little islets, all of which are clothed with luxuriant vegetation. Native prahus, with their yellow mat-sails, are occasionally seen to shoot from behind one of them, to be shielded from view immediately afterwards by the green foliage of another; and over the tops of the trees may often be descried the white sails of some stately ship, threading the mazes of this little archipelago. One group, appropriately named the Thousand Isles, has never yet been explored, and its intricacies afford concealment to petty pirates who prey upon the small prahus and

22   2 F

fishing-boats. * * * A number of large fishing-
boats were coming in from sea, and standing with
us into the roads; and although we were running
at the rate of seven knots an hour, they passed us
with great rapidity. They had a most graceful
appearance; many of them were fourteen or fifteen
tons' burthen, and each boat carried one immense
square-sail. As the breeze was strong, a thi.k
plank was thrust out to windward for an outrigger,
on which several of the numerous crew sat, or stood,
to prevent the press of sail they were carrying from
capsizing the boat. They were occasionally hidden
from our view by their passing behind some of the
small islets; but in a few seconds they would appear
on the other side, having shot past so rapidly, that
we could scarcely fancy we had lost sight of them at
all."*

In sailing amongst the numberless islands of the
Indian Archipelago, the voyager is struck with the
frequent appearance of towns or villages built
actually over the sea. The houses are constructed on
stout piles, which are firmly driven into the ground.
A flat place is selected, where the tide ebbs and
flows, that all dirt and filth from their habitations
may be regularly carried away without trouble, and
that they may be free from the presence of unpleasant and venomous reptiles. The houses are chiefly
of split bamboo, thatched with leaves: the windows
are made of the transparent inner shell of the pearl-
oyster: they are arranged in rows or streets, with
walks three or four feet wide reaching to the land,

* Earl's " Eastern Seas," p. 11.

CHINESE JUNKS.

but all heavy goods are transported by canoes, which pass under the houses. The mode of driving the piles, which are inserted into the bottom to the depth of six feet, is curious and ingenious. A canoo loaded with stones to the weight of two or three tons is lashed on each side of a pile at high water, which, as the tide falls, are suspended from it; a heavy piece of timber is then made successively to fall upon the head, which, conjointly with the great weight of the canoes, sinks it into the bottom rapidly. Towns covering a square mile may be seen formed in this manner.

The harbours and straits are crowded during the season with Chinese junks; which fail not to strike an eye accustomed to the elegant proportions and graceful tracery of an European ship, as ludicrously monstrous. Mr. Crawfurd says, "The appearance of a Chinese junk is remarkably grotesque and singular. The deck presents the figure of a crescent. The extremities of the vessel are disproportionately high and unwieldy, conveying an idea that any sudden gust of wind would not fail to upset her. At each side of the bow there is a large white spot or circle to imitate eyes. These vessels, except before the wind, are bad sailers, and very unmanageable. They require a numerous crew to navigate them: of one of the largest size, it often takes fifty men to manage the helm alone." The high stern and bow are alike flat, the latter having nothing answering to a *cut-water*. There are from two to four masts, the main-mast being disproportionately larger than the others; each of which carries a single huge

square-sail made of mats of split bamboo, extended
by horizontal rods of bamboo, on which the sail
is rolled up when reefing is necessary. The largest,
though sometimes of twelve hundred tons, have but
one deck, but the immense hold is divided into compartments, allotted to the several adventurers and
their goods. Mr. Earl describes one which he met
with in Banca Straits, in somewhat unfavourable
style. " While wind-bound," he observes, " a Chinese
junk passed close by us. A considerable number
of the crew· were standing on the high, thatched
habitation erected on their quarter-deck, and perceiving a Chinese passenger whom we had on board,
they all hailed together to demand the state of the
markets ; but they asked so many questions at once,
that our friend became quite bewildered, and the
junk passed astern before he could decide to which
he should first reply. Even if he had spoken, the
junk-people could not have profited by his efforts,
for they continued bawling until quite out of hearing. This junk, which was about two hundred tons'
burthen, carried two immense mat-sails, with a number of small yards extending along them, giving
them the appearance of bats' wings. She passed us
quickly, on account of the current being in her
favour; but, although the breeze was strong, she
went slowly through the water, and might be deemed
little better than an unwieldy hulk."*—The inflated
ideas which the Chinese maintain of their own perfection are adverse to any improvement in these
singular structures; indeed, an attempt at innova-

* Eastern Seas, p. 129.

tion, some years ago, in their form, bringing them
nearer to the model of an European ship, was so
severely reprehended in high quarters, that it was
found prudent to desist from the indiscreet improvement. At the same time, it must be confessed,
that compared with the vessels of their immediate
neighbours, the junk, as a commercial vessel, has a
vast superiority; and in the seas which they navigate, so regular are the monsoons, that they get on
tolerably well.

Occasionally, however, they must encounter those
terrific tempests called typhoons, which are peculiar
to these seas, and which, with the hurricanes of the
opposite hemisphere, are the most furious storms

SHIP UNDER BARE POLES.

that blow. They rise with fearful rapidity, often
coming on suddenly from a calm; and before the

canvas can be secured, the gale is howling shrilly
through the spars and rigging, and the crests of
the waves are torn off, and driven in sheets of spray
across the decks. The lightning is terrible: at very
short intervals the whole space between heaven and
earth is filled with vivid flame, showing every rope
and spar in the darkest night as distinctly as in the
broadest sunshine, and then leaving the sight obscured in pitchy darkness for several seconds after
each flash; darkness the most intense and absolute;
not that of the night, but the effect of the blinding
glare upon the eye. The thunder, too, peals now
in loud, sharp, startling explosions, now in long mut-
tered growls all around the horizon. In the height
of the gale, curious electrical lights, called St. Ulmo's
fires, are seen on the projecting points of the masts
and upper spars, appearing from the deck like dim
stars. Soon after their appearance the gale abates,
and presently clears away with a rapidity equal to
that which marked its approach.

The storms are found, by carefully comparing
the directions of the wind at the same time in different places, or successively at the same place, to
blow in a vast circle around a centre: a fact of the
utmost importance, as an acquaintance with this
law will frequently enable the mariner so to determine the course of his ship, as to steer out of the
circle, and consequently out of the danger; when,
in ignorance, he might have sustained the whole
fury of the tempest. The course of the circle is the
opposite of that taken by the hands of a watch, and
is the same with that of the still more striking phe-

nomena, waterspouts. These are, perhaps, the most majestic of all those "works of the Lord, and his wonders in the deep," which they behold who "go down to the sea in ships." They frequently appear as perpendicular columns, apparently of many hundred feet in height, and three feet or more in diameter, reaching from the surface of the sea to the clouds. The edge of the pillar is perfectly clean and well-defined, and the effect has been compared to a column of frosted glass. A series of spiral

WATERSPOUTS.

lines run around it, and the whole has a rapid spiral motion, which is very apparent, though it is not always easy to determine whether it is an ascending or descending line. Generally, the body of clouds

above descend below the common level, joining the
pillar in the form of a funnel, but sometimes the
summit is invisible, from its becoming gradually
more rare. Much more constant is the presence of
a visible foot; the sea being raised in a great heap,
with a whirling and bubbling motion, the upper
part of which is lost in the mass of spray and foam
which is driven rapidly round. The column, or
columns, for there are frequently more than one,
move slowly forward with a stately and majestic
step, sometimes inclining from the perpendicular,
now becoming curved, and now taking a twisted
form. Sometimes the mass becomes more and more
transparent, and gradually vanishes; at others, it
separates, the base subsiding, and the upper portion shortening with a whirling motion, till lost in
the clouds. The pillar is not always cylindrical: a
very frequent form is that of a slender funnel depending from the sky, which sometimes retains that
appearance without alteration, or, at others, lengthens
its tube towards the sea, which at the same time
begins to boil and rise in a hill to meet it, and soon
the two unite and form a slender column, as first
described.

When these sublime appearances are viewed from
a short distance, they are attended with a rushing
noise, somewhat like the roar of a cataract. . The
phenomenon is doubtless the effect of a whirlwind,
or current of air revolving with great rapidity and
violence; and the lines which are seen, are probably
drops of water ascending in the cloudy column.
They are esteemed highly dangerous: instances have

been known, in which vessels that have been crossed by them have been instantly dismasted, and left a total wreck. It is supposed that any sudden shock will cause a rupture in the mass, and destroy it; and hence it is customary for ships to fire a cannon at such as, from their proximity of course, there is any reason to dread. They are seen in all parts of the world, but are most frequent in the Pacific and Indian Oceans.

That a Chinese junk, so clumsily rigged and so unwieldy, must be ill adapted to sustain the fury of a typhoon, or to evade the rush of a waterspout, we may well imagine, and doubtless many are wrecked from these causes. The following affecting narrative of a crew under such painful circumstances will be read with interest:—

"The dark sullen waters of the China Sea never looked less friendly nor more portentous than on the morning of the 12th of January, 1837; tempestuous weather, and a sea rising in mountains around and over the ship's side, hurled her rapidly on her passage homewards, when suddenly a wreck was discovered to the westward. The order to shorten sail was as promptly obeyed as given, and the vessel was hauled towards what was discovered to be a China junk without masts or rudder, having many persons on deck vehemently imploring assistance. The exhibition of their joy, as they beheld our approach, was of the most wild and extravagant nature; but it was doomed to be transient, the violence of the elements driving the ship swiftly past the wreck. It became necessary to put her on the other tack, a manœuvre which they construed into abandonment, and the air rang with the most agonizing shrieks of misery: hope appeared to have been rekindled at the eleventh hour, but to render despair more desperate, and death more frightful.

"The excitement on board was intense. A boat was immediately lowered, in which the hawser was placed, with a small line attached to it, as a messenger, and was thrown to the wreck for the purpose of towing her to the ship; but this intention was frustrated by the breaking of the windlass to which it was fastened. The anxiety of these unfortunate people to quit their perilous position was so great, that it became dangerous to approach them: one man, in a paroxysm of despair, jumped overboard after the hawser, as the windlass broke, in the vain hope of reaching the boat; he was an expert swimmer, but no human power could prevail against that sea; the furious Ocean mocked his efforts; he rose and sunk upon the swelling billows until nature was exhausted: he was lost in sight of his companions in misfortune and of the persons sent to their aid, without any being able to afford him relief.

"Fears were entertained for the boat and her crew, as seen from the ship contending with the violence of the element in which she floated, and a moment of doubt passed the mind as to the expediency of permitting another attempt. It was only for a moment: the piercing cries borne upon the hollow blast, fell upon the sense with such terrific horror, that indecision seemed a crime; direc-

tions were then issued to keep the boat away, and
a rope with a bowline-knot at one end, was thrown
to the junk, into which signs were made for each
man to place himself, and then plunge into the
water, whence he was dragged into the boat, and
eventually, in like manner, to the ship. Thus were
eighteen persons rescued from the very grasp of
death at a moment when every ray of hope appeared to be utterly extinguished. Their gratitude
was boundless: they almost worshipped the officers,
the crew, and the vessel; prostrated themselves,
kissed the feet of the former, and the very planks
of the latter. * * * *

"After being on board five days, we made Pulo
Aor, where we took in water, and so desirous were
those simple-hearted people of testifying their gratitude, that they would not permit the men to carry
it, but filled the casks themselves; and at parting,
knelt down and kissed each man's feet with the fer-
vour of devotion. Here we separated from seventeen men who had been nine days at sea upon a
miserable wreck, water-logged, without water to
drink, and scarcely food to eat. One of them, an
old man, died on the preceding evening from the
effects of fatigue and exhaustion; the others, I doubt
not, have long ere this time reached their homes,
and taught their friends and children to bless the
Englishmen and the English ship, which, under
Providence, snatched them from a watery grave,
and returned them to their affections."*

The principal object of commercial enterprise with

* Unit. Serv. Journ. 1837, iii. 512.

the Chinese, in their annual visits to the Oriental
Isles, and, by consequence, that which forms the
chief lading of the returning junks, is the edible
birds'-nest; the production of a species of Swallow
(*Hirundo esculenta*); of which, as it seems to be ·
an oceanic production, I shall give a short account.
For many ages the nests have been in use in China,
and it is a remarkable instance of the fictitious value
often attached by fashion to things of little moment
in themselves, but procured from a distance with
much expense, difficulty, and danger. From the
accounts of travellers, which differ much in detail,
we gather, that certain large caverns in the interior
of the island, as well as on the coast, are frequented
by immense numbers of these birds, of which there
seem to be at least two species, one being, according to many observers, smaller than a wren; the
other, according to Sir. E. Home, who dissected
some brought home by Sir Stamford Raffles, "double the size of our common swallow." M. Poivre,
who, in 1741, visited the Straits of Sunda, observed
these birds in a little island called the Little Tocque.
A party having landed to shoot green pigeons, this
gentleman, accompanied by a sailor, walked along
the beach in search of shells and jointed corals,
which were very abundant. After having walked
some distance, he was called by his companion, who
had discovered a deep cavern. M. Poivre, hastening
to the spot, found the entrance darkened by an immense cloud of small birds, pouring out in swarms.
He entered, and with ease knocked down many of
the little birds, with which he was at that time un-

acquainted. As he proceeded, he found the roof of
the cave entirely covered with small nests, shaped
"like holy-water pots." Each of the nests contained two or three eggs or young, which lay softly
on feathers, such as clothed the breast of the parents.
They were found to be glued firmly to the rock, but
having detached several, and brought them on board,
they were recognized to be the same with those
which form so valuable an article of merchandize in
China. The sailor, profiting by this information,
preserved his portion, which he afterwards sold well
at Canton. The intelligent traveller, on the other
hand, took coloured drawings of his captures, and
speculated concerning the nature of the nest. He
conjectures, that it is composed of a gluey substance
often seen floating in those seas, which he considers
to be fish spawn.

More recent accounts agree generally with this.
In a little island on the coast of Java, called the Cap,
Sir George Staunton found some caverns running
horizontally into the side of the rock, in which were
numbers of these birds'-nests. "They seemed to be
composed of fine filaments, cemented together by a
transparent viscous matter, not unlike what is left
by the foam of the sea upon stones alternately
covered by the tide, or those gelatinous animal substances found floating on every coast. The nests
adhere to each other, and to the sides of the cavern,
mostly in rows without any break or interruption.
The birds that build these nests are small grey swal-
lows, with bellies of a dirty white. They were flying
about in considerable numbers; but they were so

small, and their flight so quick, that they escaped
the shots fired at them. The same nests are said
also to be found in deep caverns at the foot of the
highest mountains in the middle of Java, and at a
distance from the sea. * * * The nests are placed
in horizontal rows at different depths, from fifty
to five hundred feet. Their value is chiefly determined by the uniform fineness and delicacy of their
texture; those that are white and transparent being
most esteemed, and fetching often in China their
weight in silver. These nests are a considerable
object of traffic among the Javanese; and many are
employed in it from their infancy. The birds, having spent near two months in preparing their nests,
lay each two eggs, which are hatched in about fifteen days. When the young birds become fledged,
it is thought time to seize upon their nests, which
is done regularly thrice a year, and is effected by
means of ladders of bamboo and reeds, by which
the people descend into the cavern: but when it
is very deep, rope ladders are preferred. This operation is attended with much danger, and several
break their necks in the attempt."*

Some of the caves on the coast of Java are only
to be reached by a perpendicular descent of many
hundred feet, on these frail ladders of cane, while
the sea rages with fury far beneath the feet. When
attained, the cavern must be explored by torchlight,
the adventurous fowler securing a precarious footing
over the damp and slippery surface of the irregular
recesses, where a false step would plunge him down

* Embassy to China, i. 287.

into the boiling surf, or impale him upon the sharp processes of the rocks. The best nests are obtained from such gloomy caves as these; for there are several qualities, the best being white, or nearly transparent, as if composed of threads of isinglass; others, which are inferior, are coarser in texture, darker in colour, streaked with blood, or mixed with feathers, or defiled with the food and ordure of the young. When procured, they are simply dried in the shade, and packed in boxes, each containing a *picul*, equal to about one hundred and thirty-three pounds. In the Chinese markets they fetch prices varying, according to the quality, from 250*l.* up to above 900*l.* sterling per *picul;* the latter price being at the rate of nearly seven pounds sterling per pound, and consequently almost equal to double the weight of the article in silver! The amount shipped from the archipelago is estimated by Mr. Crawfurd at 1818 *piculs*, 242,400lbs., worth to the sellers at the islands, 284,290*l.* In defenceless and remote situations, exposed to lawless plunder, the caverns are of little value; but in other more favourable localities, the clear profit is very great; for it is computed that the whole expense of collecting, drying, and packing, does not much exceed one-tenth part of the whole amount.

The nests are used in China, by the luxurious, in thickening rich soups; but though considered by them a great delicacy, have been but little esteemed by Europeans, who have tasted the preparations at Chinese tables. The substance of which they are composed is now generally agreed to be a sea-weed

which floats on the Indian waters, a species of
Gelidium, which can be reduced, by boiling or soak-
ing in water, almost entirely into a clear jelly. It
is probable, however, that the substance undergoes
some preparation in the stomach of the bird before
it is applied, or else that the filaments are cemented
by a glutinous saliva.

No inconsiderable part of the cargoes of the
return junks is made up of a sea-weed called *agar-
agar*, collected upon the coasts of Malacca. Boats
go out to procure it from the reefs on which it
grows, when it is well washed in the rivers, dried,
and packed in baskets. It grows in small bunches,
with long and narrow fronds resembling shreds, of
a light-yellow hue. The finest portions are used
in China to make a clear, tasteless jelly; while the
coarser parts are boiled down into a strong and substantial glue, used in the manufacture of furniture
and lacquered ware. A size is also produced from
it, for stiffening paper and silk. In Canton, this
substance produces from twenty to thirty-five shillings per hundredweight. It is, however, light in
proportion to its bulk. It is probable that this is
the species described by botanists by the name of
Gracillaria tenax, of which 27,000 pounds are said
to be annually imported into China, and of which
windows are made.

Another important article of traffic with the Chinese, is the animal called by them *trepang*, the bêche
de mer (*Holuthuria*). There are several species of
these animals, which are curious creatures. Generally, they have some resemblance in form to a

cucumber, whence they are sometimes termed Sea-cucumbers; in the water, however, the body is often

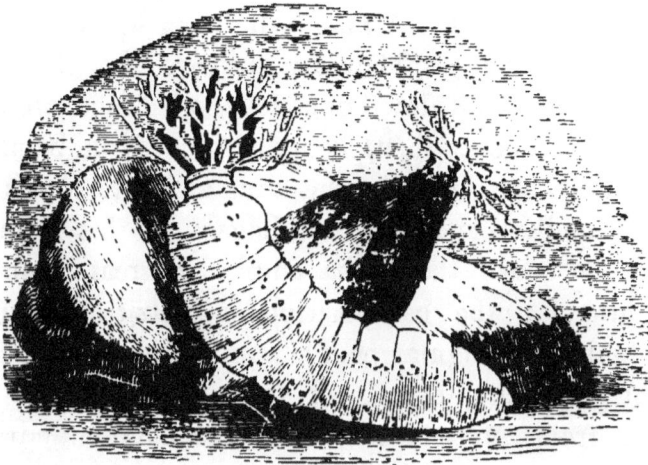

SEA-CUCUMBERS (*Holothuriæ*).

greatly lengthened, and, on being touched, is suddenly contracted so as completely to alter the form. The mouth is at one end of the animal, furnished with shelly teeth converging to a centre, as in the Star-fishes, and surrounded by numerous tentacles. Mr. Crawfurd describes it as "an unseemly-looking substance, of a dirty-brown colour, hard, rigid, scarcely possessing any power of locomotion, nor appearance of animation." The usual length is eight or nine inches, the diameter about an inch, but some are two feet in length, and seven or eight inches in girth. They frequent the shallow waters, on reefs and in lagoons; often exposed on the rock, but sometimes nearly buried in the coral-sand, their feathered tentacles alone appearing and floating loosely in the water. The large kinds are often obtained by spearing them upon the rocks in shallow water; but the ordinary mode of obtaining them is by diving in from three to five fathoms, and collecting them by hand. A man will bring up thus eight or ten at a time. They are prepared for the market by being split down one side, boiled, and pressed flat with stones: then, being stretched on bamboo slips, they are dried in the sun, and afterwards in smoke, and packed away in bags. In this state it is put on board the junks, and is in great demand in China for the composition of nutritious soups, in which that singular people so much delight. The quantity of this article of food, annually sent to China from Macassar, amounts to 8333 hundredweight; the price of which varies, according to the quality, (for there are upwards of thirty varieties distinguished in the market,) from thirty shillings sterling to upwards of twenty guineas per hundredweight. The extent of the traffic may be inferred from the number of vessels employed in it: Captain Flinders was informed, when near the north coast of New Holland, that a fleet of sixty proas, carrying a thousand men, had left Macassar for that coast two months before, in search of this sea-slug; and Captain King was informed that two hundred proas annually leave Macassar for this fishery. They sail in January, coasting from island to island, till they reach Timor, and thence steer for New Holland, when they scatter themselves in small fleets, and having fished along the coast, return about the end of May, when the westerly monsoon breaks up.

The periodical change of the direction of the

wind in the northern part of the Indian Ocean, by which the north-east trade-wind is exchanged for one directly opposite, commonly called the setting in of the south-west monsoon, is attended with very remarkable effects on the weather. It is the commencement of the rainy season, which is ushered in by storms of thunder, lightning, and rain, of such violence, as those acquainted only with a temperate climate have no conception of. Mr. Elphinstone thus describes the scene on the coast of India: "The approach of the monsoon is announced by vast masses of clouds that rise from the Indian Ocean, and advance towards the north-east, gathering and thickening as they approach the land. After some threatening days, the sky assumes a troubled appearance in the evenings, and the monsoon in general sets in during the night. It is attended by such a thunderstorm as can hardly be imagined by those who have only seen that phenomenon in a temperate climate. It generally begins with violent blasts of wind, which are succeeded by floods of rain. For some hours lightning is seen almost without intermission; sometimes it only illumines the sky, and shows the clouds near the horizon; at other times it discovers the distant hills, and again leaves all in darkness: when in an instant, it reappears in vivid and successive flashes, and exhibits the nearest objects in the brightness of day. During all this time the distant thunder never ceases to roll, and is only silenced by some nearer peal, which bursts on the ear with such a sudden and tremendous crash as can scarcely fail to strike the most insensible heart with awe. At length the thunder ceases, and nothing is heard but the continued pouring of the rain, and the rushing of rising streams. The next day presents a gloomy spectacle: the rain still descends in torrents, and scarcely allows a view of the blackened fields; the rivers are swollen and discoloured, and sweep down along with them the hedges, the huts, and the remains of the cultivation which was carried on during the dry season in their beds."*

The effect upon the sea is graphically depicted by Mr. Forbes: "At Anjengo," observes this author, "the monsoon commences with great severity, and presents an awful spectacle: the inclement weather continues with more or less violence, from May to October. During that period the tempestuous Ocean rolls from a black horizon, literally of 'darkness visible,' a series of floating mountains heaving under hoary summits, until they approach the shore; when their stupendous accumulations flow in successive surges, and break upon the beach; every ninth wave is observed to be generally more tremendous than the rest, and threatens to overwhelm the settlement. The noise of these billows equals that of the loudest cannon, and with the thunder and lightning so frequent in the rainy season, is truly awful. During the tedious monsoon I passed at Anjengo, I often stood upon the trembling sand-bank to contemplate the solemn scene, and derive a comfort from that sublime and omnipotent decree, 'Hitherto shalt thou come, but no further; and, here shall thy proud waves be stayed!'"†

* Account of Caubul, p. 126.   † Oriental Memoirs.

An effect, scarcely less sublimely magnificent, is produced by the coming in of the periodical spring-tide at the mouth of some of the large rivers of India, which is called the Bore. The rising flood, confined by the narrowing coasts of a deep estuary, takes the form of an immense wave, which comes majestically rolling along, like an advancing cataract, bearing every thing before it. So rapid is its march, that its progress from Hooghly Point to Hooghly Town, a distance of seventy miles, occupies but four hours. At Calcutta the wave is five feet high; but in the channels formed by the numerous islands in the Burhampooter, its height is twelve feet; and so terrific is it, that no boat dares to navigate the river at the time of spring-tide. As the middle of the river, however, is comparatively free from the influence, and only one side, usually, is subject to its greatest violence, the boats and larger craft hasten, on its approach, into the open water of the current; but if unhappily overtaken, they are inevitably overturned or swamped, while even large ships, that present their broadsides to its advance, are rolled so violently, that their yard-arms are dipped in the wave.

The multitudes of fishes of brilliant hues and fantastic shapes, that play in the tepid waters of these regions of the sun, are incalculable. Numerous bands of Parrot-fishes (*Scarus*) and Rock-wrasses (*Labrus*) sport about the reefs, whose bodies are ornamented with crimson, yellow, and silvery tints, often arranged in the form of bands or stripes; Gurnards (*Trigla*), whose large fins resemble in their

form and delicate pencillings the wings of a butterfly, take momentary flights above the surface; and
the petty tribe of *Chætodons*, several of which are
noted for the singular habit of shooting flies with
a drop of water projected from their beak-like
mouths, fearlessly approach the hand immersed in
the water. But none of these are more curious than
the Toad-fishes, or Anglers (*Antennarius*), whose
pectoral and ventral fins have much of the form and
also the functions of the feet of a quadruped, enabling them to crawl out of the water, and travel
over the land. The head is armed with horn-like
projections, terminating in shining filaments, which
play freely in the water, and attract small fishes
within the reach of its enormous mouth; a very
remarkable instance of the superintending care exercised by the beneficent Creator over the well-being of
his creatures. The form of the fish is clumsy, and
its motions slow and heavy, and without this provision for the attraction of its prey, it would probably
fare but poorly.

It is doubtless a species of *Antennarius* that is
thus described by Mr. Earl, as observed on the coast
of Borneo: "Large tracts of mud had been left
uncovered by the receding tide, and flocks of gulls
and other birds were feeding on the worms and small
fish. Vast numbers of little amphibious creatures
were running about in the mud, and they appeared
to be sought after by some of the larger birds.
They were from two to eight inches long, resembling a fish in shape, of a light-brown colour, and
could run and jump by means of two strong pectoral

fins. On the approach of an enemy, they buried themselves in the mud with inconceivable rapidity, so that their sudden disappearance seemed to be the work of magic. One of the Malays was employed in catching them, as they are considered to be a great delicacy. He used for the purpose a thin plank, four feet long, and one foot broad; on one end of which were fixed several sharp-pointed nails, the points projecting beyond the end of the plank. He placed the plank flat upon the mud, and with the right knee resting on it, and kicking the mud with the left foot, he shot along the surface with great rapidity, the sharp-pointed nails transfixing the little creatures before they could succeed in burying themselves sufficiently deep to avoid it. This is a dangerous sport, and requires great skill in the fishermen to prevent accidents; for should he lose his plank, death would be almost inevitable, the mud not having sufficient consistence to support him without the aid of this simple contrivance."*

Numberless creatures of the inferior classes, some of which are of exquisite delicacy and beauty, float on the surface of the Indian Ocean; often in such immense hosts as to cover the sea for miles around. The Violet-snail (*Janthina fragilis*) is one of these, whose shell much resembles that of our garden-snail in form and size, but is of a pearly-white above, and beneath violet. When alive it is covered with a slippery membrane. A singular floating apparatus projects horizontally from the aperture of the

* Eastern Seas, p. 213.

shell, resembling a collection of air-bubbles, but composed of a delicate white membrane, inflated, and puckered on the surface into the bubble-like divisions alluded to; it is oblong, about an inch in length. The buoyancy of this float supports the animal at the surface, where it lies with the convexity of the shell downward. Three or four drops of a blue liquid are contained in the body, which has been supposed to answer the purpose of concealment in time of danger, by imparting an obscurity to the water; but it is hardly sufficient for this purpose, as the whole quantity secreted by one animal will not discolour half a pint of water. Beneath the float, at certain seasons, the eggs are suspended by pearly threads; and as the floats are frequently found in great numbers with eggs thus attached, but separate from the original animals, it is thought that they have the power of throwing off this appendage and forming a new one; in which case it serves the purpose of sustaining the eggs, and probably the young, within the reach of the light and heat of the sun.

The Portuguese Man-of-war (*Physalis pelagica*), numerous in the warm parts of the Atlantic, is still more abundant in the seas of which I am writing. It is a beautiful little creature, though of very simple structure, consisting merely of a semi-transparent membranous bag, round at one end, and pointed at the other, along one side of which runs a wide membrane, puckered into perpendicular folds, and capable of being contracted and dilated; while from the opposite side depends a thick fringe of blue tentacles, among which are some of a great length, and of a crimson and purple hue. The tentacles have the faculty of severely stinging the hand that touches them, though ever so slightly; and it is probable that this power is in some way connected with the sustenance of the animal, as minute fishes are frequently found in a benumbed state attached to these processes. The little creature, as it floats upon the broad billows, bears a very striking resemblance to a little ship, of which the bladder is the hull, and the puckered membrane the sail; and as the edge of the sail is a beautiful pink hue, and the lower part of the hull deep blue, a fleet of them, floating and rolling in a calm upon the long glassy swell of the sea, presents a scene of striking novelty and elegance.

Another creature much resembling this in appearance is found in the same regions in equal numbers. It is called by sailors the Sallee-man (*Velella mutica*); and consists of an internal cartilage, of a semi-pellucid white hue, enclosed in soft parts, of a purplish green. A broad oval base floats on the water, across which runs obliquely an arched crest or sail: beneath are placed the brown viscera, covered with a thick mat of colourless tubular *papillæ:* the edge of the oval base is fringed with slender blue tentacles. No part of this animal seems to have the power of stinging, so formidable in the preceding.

It will be remembered, that in the description of the Arctic Seas, a little animal (*Clio borealis*) was mentioned as forming a large portion of the food of the whale. Its place is supplied in the Pacific

and Indian Oceans by two or three species nearly
allied to it in structure, but furnished with a glassy
shell.   One of these is named *Hyalea tridentata;*

GLASS SHELLS.   (*Hyalea tridentata, and Cleodora pyramidata.*)

its shell is small and somewhat globular, resembling
a bivalve without a hinge; the hinder part being
consolidated and armed with three spines; the sides
have a narrow fissure through which a semi-transparent membrane protrudes.   The animal is furnished with a wing or fin on each side, which it uses
as oars.   A kindred species (*Cleodora cuspidata*) is
of extreme delicacy and beauty.   The shell is glassy
and colourless, very fragile, nearly in the form of
a triangular pyramid, with an aperture at its base,
from which proceeds a long and slender glassy spine;
and a similar spine projects from each side of the
middle of the shell.   The animal is like the preceding; but the hinder part is globular and pellucid,
and in the dark vividly luminous, presenting a sin-
gularly-striking appearance, as it shines through its
perfectly-transparent lantern.   Both of these are
found floating in great numbers on the surface of
the sea.

Among the sea-shells which attain a large size in these seas, the Giant Clamp (*Tridacne gigas*) stands pre-eminent. It is found in abundance on the coasts of Sumatra, as well as of other islands, attached to the rocks by a strong cable. This, which is called byssus, is formed of many tough threads, but slightly elastic, spun by the animal, or rather, *cast in a mould* thread by thread; a glutinous fluid being secreted in a long groove or canal formed by the foot, which in the air rapidly acquires solidity. When complete, the united threads form, as observed above, a cable, projecting through an opening in the back of the shell, and adhering by the other extremity to the rock, so firmly as to resist the agitation of the sea, and so tough as to be severed only by an axe. Marsden mentions one which was more than three feet three inches long and two feet one inch wide: and specimens have been seen which had attained the enormous length of four feet. They are sometimes taken, when not adhering, by thrusting a long bamboo between the open valves, which immediately close firmly, and they are dragged out. The substance of the shell is perfectly white, several inches thick; and is worked by the natives into arm-rings, and by European artists is made to receive a polish equal to the finest statuary marble.

Pearls, whose exquisite beauty have made them celebrated from the earliest ages, are well known to be marine productions; and as the shores of the Indian Ocean yield the finest specimens, I may here say a word of the fishery for them. Many bivalve shells produce pearls of greater or less perfection; but what is known as the Pearl Oyster is the *Avicula margaritifera* of conchologists. The interior surface of the shell is covered with very thin plates, or *lamellæ*, which are furrowed with microscopically minute and close parallel grooves, and in this structure lies the property of reflecting opaline tints; a property which has been communicated to other substances by mechanically impressing the surface with similar grooves. In some diseased states of the animal, or when the shell has received a trifling injury, or some foreign body—a grain of sand, for example—has found its way within the mantle, the pearly secretion is poured out in great abundance around the part, and, layer being imposed upon layer, produces a pearl, either attached to the inner surface of the shell, or loose and held merely in the folds of the mantle.

The most productive fishery is in the Persian Gulf, and the finest pearls are found there; above 90,000*l.* sterling are sometimes realized from this source in the course of two months. Those with which we are most acquainted, are carried on on the coasts of Coromandel and of Ceylon; the former being in the hands of the East India Company, the latter in those of the British Government. The Ceylon fishery has been well described by Captain Percival, the Count de Noe, and lately by Mr. Bennett. As the banks would soon be exhausted if fished every year, portions only are selected in turn, while the rest remains untouched to be recruited. In the month of November, the Government appoints an inspection of the state of the banks, and those selected as fit for fishing are advertised accordingly, the fishery for the ensuing season being offered for sale. In January, the boats begin to assemble, and the adventurers from all parts of India congregate on a narrow spot of barren sand which is deserted for the greatest portion of the year, but now presents the life and gaiety of a fair. "There is, perhaps, no spectacle," says Captain Percival, "which the Island of Ceylon affords, more striking to an European than the bay of Condatchy during the season of the pearl-fishery. This desert and barren spot is at that time converted into a scene which exceeds in novelty and variety almost any thing I ever witnessed; several thousands of people of different colours, countries, castes, and occupations, continually passing and repassing in a busy crowd; the vast numbers of small tents and huts erected on the shore, with the bazaar or market-place before each; the multitude of boats returning in the afternoon from the pearl banks, some of them laden with riches; the anxious expecting countenances of the boat-owners, while the boats are approaching the shore, and the eagerness and avidity with which they run to them when arrived, in hopes of a rich cargo; the vast numbers of jewellers, brokers, merchants, of all colours, and all descriptions, both natives and foreigners, who are occupied in some way or other with the pearls, some separating and assorting them, others weighing and ascertaining their number and value, while others are hawking them about, or drilling and boring them for future use;—all these circumstances tend to impress the mind with the value and importance of that object which can of itself create this scene."*

The actual fishery begins in February and con-tinues during six weeks, or at most two months. The boats, being prepared, each carrying twelve or fourteen hands and ten divers, leave the shore at the signal-gun of the government officer, and arrive at the bank before daylight. At sunrise diving com-mences, and the divers, divided into two parties, descend alternately, the one set breathing while the other is below. To expedite his descent, each man has a conical piece of granite, through a hole in which a rope is passed; he grasps the rope with the toes of his right foot, which he uses with nearly the same pliancy as the fingers of his hands, and taking in his left a net like an angler's landing-net, seizes another rope in his right hand, and closes his nostrils with his left thumb and finger. The weight of the stone causes him to descend rapidly, and he loses no time, but hastily fills his net with the oys-ters he finds around. When he can retain his breath no longer, he jerks the second rope, and is instantly hauled to the surface by his fellows, leaving the stone to be pulled up afterwards. Generally, from a minute and a half to two minutes, is as long as a diver can remain under water; but Captain Per-cival records a case in which a man "absolutely re-mained under water full six minutes." The effects of so long a submersion as even ordinarily takes place, are severe, and manifest themselves by gush-ings of water from the ears, mouth, and nose, and sometimes by discharges of blood. Yet they are

* Percival's Ceylon, p. 59.

ready to take their turn again, frequently making
forty or fifty plunges a day, and bringing up at each
turn about a hundred oysters.

The greatest danger to these adventurous men
arises from the sharks, to whose rapacity allusion
has before been made. But against them the poor
people believe that they possess an inviolable defence in the charms sold to them by pretended con-
jurors, whose impudence and address secure their
hold on their deluded votaries, even in spite of the
frequent evidence of their fallibility. It is probable,
the constant bustle and noise, and the frequent
splashings of the divers, deter the sharks in a great
measure from approaching the scene.

" As soon as the oysters are landed, they are placed
in pits on the shore, and left to undergo decomposition; in which state they diffuse an intolerable odour,
but to which habit speedily reconciles the people.
When the flesh is decayed under that burning sun,
the shells are opened with ease, and minutely examined for pearls: some, however, elude the utmost
vigilance, to obtain which, numbers of people continue
to search the sands for months after the merchants
have departed, and they are now and then rewarded
by a pearl of value. In 1797, a common fellow, of
the lowest class, thus got by accident the most
valuable pearl seen that season, and sold it for a
large sum."

In the Straits of Sunda and the adjacent seas,
there are found several floating sea-weeds, which
have a general resemblance to the Gulf-weed of the
Atlantic, but possess a much more striking similarity

24

to terrestrial plants. Two species in particular, named from this resemblance *Sargassum aquifolium* and *S. ilicifolium*, so closely imitate our common holly in their branches, berries, and twisted spinous leaves, as to induce a belief, at the first glance, that they are no other than sprigs of that familiar plant. Another species, found in the same locality, is called *S. Taxifolium*, from its likeness to the yew. The former are highly interesting on another account: they afford a remarkable illustration of the fact, that the seed-receptacles of some sea-plants are metamorphosed after the discharge of their seeds into leaves and air-vessels. Few would suspect that the round air-cells, that look like green berries, or the curled and thorny leaves, were alike the slender pro- cesses containing the seed, only in another stage of development; yet specimens are often found in which the process is actually going on, both the one and the other being but partially transformed. The pores with which the surface of the leaves are stud- ded, are but the orifices through which the seeds escaped.

As we approach the Cape of Good Hope, the sea- birds peculiar to high latitudes again appear, and the sea and air are enlivened by myriads of gulls, terns, petrels, frigate-birds, and albatrosses. But among them we have yet to notice one pre-eminent among them, a master-fisher, which, for its powers of consuming the finny prey, is perhaps unrivalled. It is the Pelican (*Pelicanus onocrotalus*), which abounds all around the shores of the Indian Ocean, ranging to the distance of several hundred miles from the coasts. This bird has great powers of flight, the extended wings covering a space of twelve feet. The throat is dilated into a capacious bag, which can be wrinkled up when not in use, but when the animal is fishing forms a convenient pouch, in which the prey is stored as it is caught, until it is filled, when the booty is borne to shore, to feed the callow young, or to be eaten at leisure. The pouch of a full-grown Pelican, when distended, will contain ten quarts of water. They fly to a long distance, and at a lofty elevation, and remain untired on the wing for a protracted period. A flock of Pelicans beating for prey is a splendid spectacle. Sometimes the whole troop soars upwards to an immense height, and then suddenly swoops down with arrowy velocity, splashing the sea in every direction; presently they emerge, and again soar on high, till again they simultaneously dash down upon the shoals; and thus the flock perform their evolutions in concert, ranging over a wide bay, or a given space of water, with perfect order and regularity, and with astonishing rapidity. At other times they fly almost at the very surface, beating the water with their wings, till the whole sea is one undistinguishable mass of foam.

In the beautiful poem of Montgomery, "The Pelican Island," which I have before quoted, the manners of these interesting birds are ably described :—

> " Eager for food, their searching eyes they fix'd
> On ocean's unroll'd volume, from a height
> That brought immensity within their scope;
> Yet with such power of vision look'd they down,

As though they watch'd the shell-fish slowly gliding
O'er sunken rocks, or climbing trees of coral.
On indefatigable wing upheld,
Breath, pulse, existence, seem'd suspended in them:
They were as pictures painted on the sky;
Till, suddenly, aslant, away they shot,
Like meteors changed from stars to gleams of lightning,
And struck upon the deep; where, in wild play,
Their quarry flounder'd, unsuspecting harm;
With terrible voracity, they plunged
Their heads among th' affrighted shoals, and beat
A tempest on the surges, with their wings,
Till flashing clouds of foam and spray conceal'd them.
Nimbly they seized and secreted their prey,
Alive and wriggling in the elastic net,
Which Nature hung beneath their grasping beaks;
Till swoll'n with captures, the unwieldy burthen
Clogg'd their slow flight, as heavily to land
These mighty hunters of the deep return'd.'
There on the cragged cliffs they perch'd at ease,
Gorging their hapless victims one by one;
Then, full and weary, side by side they slept,
Till evening roused them to the chase again."

I have reserved till the last of these gleanings from
the Ocean, one of the most curious of its phenomena,
and one that, while it vividly strikes the fancy of the
voyager when he beholds it for the first time, fails
not to maintain its power to interest after years of
observation have made it familiar. I have reserved it
until the last, because it is peculiar to no sea, but
common to all, being observable in the frozen ocean
of either pole, and under the burning line; in the
Atlantic and in the Pacific. Still there seem to be
greater intensity and brilliance in the display of the
phenomenon in the tropical seas than in colder
climates. No sooner has night descended over the

Ocean, than the whole surface is seen to be, as it were, composed of light, assuming, however, various forms and aspects. The most usual appearances, as far as they have fallen under my own observation in the Atlantic, are as follows: On looking over the stern, when the ship has steerage-way, her track is visible by a line or belt of light, not a bright glare, but a soft, subdued yellowish light, which immediately under the eye resembles milk, or looks as though the keel stirred up a sediment of chalk which diffuses itself in opaque clouds through the neighbouring water, only that it is light and not whiteness. Scattered about this cloudiness, and particularly where the water whirls and eddies with the motion of the rudder, are seen innumerable sparks of light distinctly traced above the mass by their brilliancy, some of which vanish and others appear, while others seem to remain visible for some time. Generally speaking, both these phenomena are excited by the action of the vessel through the waves, though a few sparks may be observed on the surface of the waves around. But now and then, when a short sea is running without breaking waves, there are seen broad flashes of light from the surface of a wave, coming and going like sudden fitful flashes of lightning. These may be traced as far as the sight can reach, and in their intermittent gleams are very beautiful : they have no connection with the motion of the ship. In a voyage to the Gulf of Mexico, I saw the water in those seas more splendidly luminous than I had ever observed before. It was indeed a magnificent sight, to stand in the fore part of the vessel and

2 I

watch her breasting the waves. The mass of water
rolled from her bows as white as milk, studded with
those innumerable sparkles of blue light. The
nebulosity instantly separated into small masses,
curdled like the clouds of marble, leaving the water
between of its own clear blackness; the clouds soon
subsided, but the sparks remained. Sometimes one
of these points, of greater size and brilliancy than
the rest, would suddenly burst into a small cloud of
superior whiteness to the mass, and to be then lost
in it. The curdling of the milky appearance into
clouds and masses, and its quick subsidence, were
what I had never observed elsewhere.

Many very interesting observations have been made
on these luminous appearances, and there seems no
doubt that to a very large extent they are produced
by living animals; but as many species, varying
greatly from each other, and belonging even to different classes of the animal kingdom, have been recognized as contributing to the luminousness, we need
the less wonder that there should be variations in its
aspects. Dr. Baird, in some quotations from a journal kept during a voyage to India, furnishes some
interesting notes of the origin of the light. The
writer speaks of "the broad bright flash, vivid enough
to illuminate the sea for some distance round, while
the most splendid globes of fire were seen wheeling
and careering in the midst of it, and by their brilliancy outshining the general light." On drawing a
bucket-full of water the narrator "allowed it to remain quiet for some time, when, upon looking into it
in a dark place, the animals could be distinctly seen

emitting a bright speck of light. Sometimes this was like a sudden flash, at others appearing like an oblong or round luminous point, which continued bright for a short time, like a lamp lit beneath the water, and moving through it, still possessing its definite shape, and then suddenly disappearing. When the bucket was sharply struck on the outside, there would appear at once a great number of these luminous bodies, which retained their brilliant appearance for a few seconds, and then all was dark again. They evidently appeared to have it under their own will, giving out their light frequently at various depths in the water, without any agitation being given to the bucket. At times might be seen minute but pretty bright specks of light, darting across a piece of water, and then vanishing; the motion of the light being exactly that of the *Cyclops* through the water. Upon removing a tumbler-full from the bucket, and taking it to the light, a number of *Cyclops* were accordingly found swimming and darting about in it."* Dr. Baird concludes from these facts that the bright globes were large Sea-blubbers (*Medusa*), and that the sparks were minute *Entomostruca*, somewhat similar in form to those figured in the former part of this volume.

In some highly interesting observations made during a series of years by M. Ehrenberg, chiefly in the Red Sea, we find many minute animals mentioned as luminous; but it is remarkable that after many trials he could not detect the slightest light from any species of the *Entomostraca*. The water was found

* Zoologist, 1843, p. 55.

to be very full of small slimy particles without any
definite form, which gave out light when the water
was stirred. These were probably *Medusæ*, torn but
yet living, as in some cases fragments of these animals are very tenacious of life. Several minute *Medusæ* of various species gave out light, which seemed
to be more vivid on any extraordinary excitement of
the animals. A drop of sulphuric acid being put
into a glass of water, several bright flashes of light
were seen. One of the little animals was taken up in
a drop of water on the point of a pen; on a drop of
acid being added, it gave out a momentary spark and
instantly died. Several new species of luminous
animals were discovered by thus mingling acid with
quantities of sea-water. The light of different species is found to vary in character; some of the sparks
being yellow and dull, others clearer and whiter, and
more lasting. The creature which produces the
brightest light of all is a kind of sea-worm (*Nereis
cirrigera*); it lives in groups or large masses, among
the branches of sea-weed; and when portions of this
are thrown on shore by the waves, the animals sur-
vive and continue to shine very brilliantly for several
days. In our own seas, a great deal of the light is
owing to the presence of an exceedingly minute
animal (*Noctiluca miliaris*), which does not exceed
$\frac{1}{1000}$ part of an inch in diameter. It consists of a
transparent globe, with a kind of tail proceeding
from one part of the circumference. In the interior
may be seen an oval nucleus, not in the centre, from
which proceed numerous branching vessels. The
luminous property appears to reside in these vessels

which, while the animal is alive, are seen to dilate and contract with a very rapid pulsation. The little globe is propelled in any direction by a jerking mo-

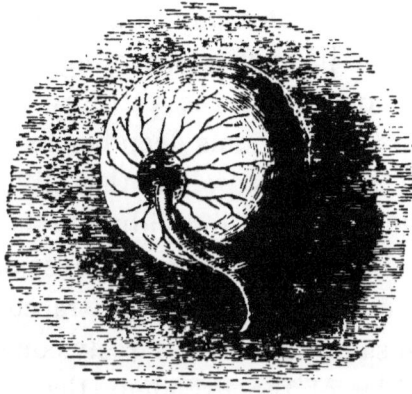

Noctiluca Miliaris, greatly magnified.

tion of the tail or stem; and as it is a restless creature, it is not a very easy matter to obtain a good sight of it for observation.

Several species of fishes are undoubtedly luminous: the Sun-fish (*Cephalus mola*), when seen at a considerable distance below the surface in a dark night, is said to glow like a cannon-ball heated to whiteness. Ehrenberg found that the whole skeleton of an Egyptian fish (*Heterotis Nilotica*) emitted such a vivid light as he never saw equalled by any other fish, alive or dead. And Mr. F. D. Bennett discovered a new species of Shark, which he named *Squalus fulgens*, from the whole surface of whose body proceeded a greenish light, which rendered the animal the most ghastly object imaginable. But there can be no doubt that the main source of oceanic efful-

gence is to be found in the countless millions of
minute animals which throng the sea, but which are
invisible without the aid of high microscopic powers.
And, truly, when from a lofty station on board a ship
we survey a space of many square miles, and see
every portion of its surface gleaming and flashing in
living light; or mark the pathway of the vessel
ploughing up from fathoms deep her radiant furrow,
so filled with luminous points that, like the milky
way in the heavens, all individuality is lost in the
general blaze, and reflect that wherever on the broad
sea that furrow happened to be traced, the result
would be the same; one can scarcely conceive a more
magnificent idea of the grandeur, the unimaginable
immensity of the Creation of God.

MAN'S SUBMARINE WORK.

Notwithstanding the great amount of information which has been gathered, through past ages,
concerning the ocean and its treasures, the sum total
compared with the vast store as yet untouched is
almost trifling. The present century, so preëminent
for progress in all departments of human investigation, has also made great advancement in ocean lore.
The diving-bell with its manifold improvements,
has enabled us to literally walk upon the bottom of
the sea, gather, at comparative leisure, the wealth
of the mines of the great deep, and examine its caverns and treasure-houses with considerable security.
While formerly, only the shores and edges of the
wide waters could be searched, now, the diver with
his water-proof armor boldly plunges into the waves,
wherever the hope of reward may present itself.

The simplest form of diving apparatus, the diving
bell, resembles a huge bell. It may be square,
or irregular. It is simply an air-tight box open at
one end. Seats are arranged on the inside for the
convenience of those who go down in it. This box
is put on the water, open end down, and the diver
enters through a small door in the top or side. He
closes the door, which is perfectly air and water-
tight,—gives the signal to those outside who are to

manage the " bell," and they lower it into the water slowly until it reaches the bottom or until the
diver makes a signal to stop. The diving bell sinks
by its own weight. As it goes down, water cannot
get into it because it is full of air which cannot
escape. The air however becomes very dense and, at
great depths, very oppressive.

When the bell touches the bottom, the diver gets
off the seat and works around inside the bell at
whatever is undertaken. He can have a light with
him, and various tools, and can labor for a short
time, as though not under water. The air however
soon becomes vitiated by his breathing and by the
burning of his light, so that he would suffocate after
a while, unless fresh air was supplied. Furnishing
fresh air by the means of a pump and tube connecting with the bell was one of the first improvements.

It has been found, in practice, however, that in
the very best form the diving-bell is a cumbersome,
clumsy affair. The latest and most approved model
is constructed in three compartments, one of which
is filled with compressed air, and one designed to
be filled with water, to assist in sinking the machine
while the whole apparatus is under the control of
the diver. But he is limited in his work to the
small space bounded by the sides of the bell. As this
heavy sub-marine house cannot be moved, in a lateral direction, without great trouble and as the movements of the diver is so closely circumscribed as
to render him useless in many kinds of submarine
work, a sort of water-proof armor has been inven-
ted which seems to answer every purpose.

At first this armor was made of cloth impervious to water, but it soon wore out or became punctured by sharp rocks or nails about a wreck so as to admit water, and become useless. The discovery of India-rubber and the great improvements in working iron and steel have made it possible to construct a diving armor which is nearly perfect. It can be put on in a few minutes—the flexible joints allowing great freedom to the diver. The mask for the face and head can be opened and closed in a moment, and the diver carries down with him a supply of air in a light steel chest or knapsack on his shoulders. This supply is kept fresh and full, by means of a pipe communicating with a pump above. The diver takes between his lips a tube connecting with the air chamber, and breathes as naturally as if on land. The air which escapes from his lungs at each expiration rises to the surface in bubbles just as regular as his breathing, giving notice to those above that the diver is alive. Should the air bubbles cease to rise, that would be evidence that he had stopped breathing, and he would be instantly drawn up to the surface by means of a cord which is attached to him for that purpose. Should the breathing as shown by the bubbles be irregular, then it would be evidence that the diver was in some trouble.

The steel air-chamber is admirably contrived for the purpose. No matter how fast the pump above is operated, it does not affect the diver. He can draw from his air-reservoir at his convenience, and when he stops drawing, no air can escape, for a valve is

arranged to prevent it, only as wanted by the diver.
Equipped with his ingenious paraphernalia the diver
roams about the bottom of the ocean almost at pleasure.   He has great glass eyes in his mask, which
covers his head and face, and through them he can
see quite plainly unless the water is very muddy or
very deep. In such cases the diver is provided with
an electric lamp which burns under water and gives
a brilliant light.   He can enter the cabin of a sunken ship, gather up the treasures found there, and
bring them to the surface.   When the articles are
too heavy to permit that, he fastens on the tackling
necessary to raise them by machinery.

In exploring sunken wrecks, in visiting the holds
and cabins of vessels in search of valuables, the
diver frequently meets with strange adventures,
and views scenes fearful enough to make the stout-
est heart quail.   The bodies of the drowned, are
sometimes disturbed by the agitation of the water,
as he moves about, and come towards him as if to
clutch him in their slimy embrace. He pushes them
one side, only to have them return to him again
with perhaps others which are caused to float by
the commotion. Sometimes he is attacked by ferocious sea monsters and is obliged to stop work and
defend himself, as best he can, but these incidents
do not hinder the brave men in the prosecution of
their work. The uses of the diving armor are constantly multiplied.   The bottom of vessels are now
examined, scraped and mended without being put
on the dry dock.   The foundations for bridges are
surveyed and numerous other purposes subserved

which readily suggest themselves to careful readers.

But, there is a limit to the field of the most expert diver, and with the best of preparation. As he descends in the water, the pressure upon him constantly increases, it is estimated, at the rate of one atmosphere for every thirty-two feet of depth, so when he is down thirty-two feet, he is subjected to a pressure of two atmospheres; at sixty-four feet the pressure is three atmospheres and so on. When he reaches a depth of two hunderd feet the pressure is so great that he becomes dizzy, the blood gushes from his nostrils and he becomes insensible. It is found that the depth of one hundred and eighty feet is the lowest in which a diver can operate with success.

The ocean has always been a formidable barrier obstructing man's progress in subduing Earth, but modern thought, industry and science are daily overcoming difficulties seemingly insurmountable.

Human ingenuity has made a pathway for the lightning through the midst of the deep, and thus, at one gigantic sweep annihilated the world of waters which so long had hindered the intercourse of nations. How simple the matter when once accomplished. A wire laid upon the bottom of the sea, and a thimble full of acid, and time and distance and oceans are as nothing.

Such vast results have followed the achievement of laying the Atlantic cable, and the public heart was so thrilled with the success, that cable and ocean have ever since seemed to be parts of each other. One bears precious friends, goods, wares and merchandise between continents isles and nations, the other carries the no less precious, but imponderable, wealth of messages from land to land.

The first attempt to lay the cable, in the year 1857, was a failure, because the wire parted and the vessel returned with a sorrowful but not disheartened company. The next effort was the following year, with a stronger cable and better machinery for "paying" it out. The cable was seven strands of wire made into a rope and covered with several thicknesses of substances designed to keep the water from reaching the metallic wires which conducted the electricity. When finished it weighed nearly two thousand pounds to the mile in length and was so strong that six miles of it could be drawn straight in water. The number of cables will increase as fast as needed.

In 1873 there were three in full operation, and they probably will last many years. It requires but a small quantity of electricity to operate them.

The signals used in ocean telegraphy differ from those used for land lines. On the French line small flashes of light are reflected on a mirror and serve the purpose. The small amount of electricity used on long cables obviates the danger of injury to the wires, by an overcharge. Science has also enabled the electrician, in case of a break in the wire, to tell almost its exact place, to which a vessel can be sent to make repairs. Thus it is that the ocean is subjected to man's dominion day by day.

THE END.

www.ingramcontent.com/pod-product-compliance
Lightning Source LLC
Chambersburg PA
CBHW021357210326
41599CB00011B/910